2/1/93

**Use and Abuse
of Statistical Methods
in the Earth Sciences**

INTERNATIONAL ASSOCIATION FOR MATHEMATICAL GEOLOGY
STUDIES IN MATHEMATICAL GEOLOGY

1. William B. Size (ed.):
*Use and Abuse of Statistical Methods
in the Earth Sciences*

Use and Abuse
of Statistical Methods
in the Earth Sciences

Edited by William B. Size
Emory University

New York * Oxford
OXFORD UNIVERSITY PRESS
1987

Oxford University Press

Oxford   New York   Toronto
Delhi   Bombay   Calcutta   Madras   Karachi
Petaling Jaya   Singapore   Hong Kong   Tokyo
Nairobi   Dar es Salaam   Cape Town
Melbourne   Auckland
and associated companies in
Beirut   Berlin   Ibadan   Nicosia

Copyright © 1987 by Oxford University Press, Inc.

Published by Oxford University Press, Inc.,
200 Madison Avenue, New York, New York 10016

Oxford is a registered trademark of Oxford University Press

Library of Congress Cataloging-in-Publication Data
Use and abuse of statistical methods in the earth sciences.
(Studies in mathematical geology; 1)
"Developed from a symposium of the same name held in
Reno, Nevada, November of 1984, at the Annual Meeting of
the Geological Society of America and sponsored by the
International Association for Mathematical Geology"—Pref.
Includes index.
1. Earth sciences—Statistical methods—Congresses.
I. Size, William B.   II. International Association for
Mathematical Geology.   III. Geological Society of America.
Meeting (1984 : Reno, Nev.)   IV. Series.
QE33.2.M3U84 1987   550'.72   86-23811
ISBN 0-19-504963-2

2 4 6 8 10 9 7 5 3

Printed in the United States of America
on acid-free paper

Foreword to the Series

This series of studies in mathematical geology provides contributions from the geomathematical community on topics of special interest in the earth sciences. As far as possible, each volume in the series will be self-contained and will deal with a specific technique of analysis. For the most part, the results of research will be emphasized. An important part of these studies will be an evaluation of the adequacy and the appropriateness of present geomathematical and geostatistical applications. It is hoped the volumes in this series will become valuable working and research tools in all facets of geology. Each volume will be issued under the auspices of the International Association for Mathematical Geology.

Richard B. McCammon
U.S. Geological Survey
Reston, Virginia

Preface

This book is meant for earth scientists who want to take a closer look at some of the underlying assumptions of randomness, normality, and variance of sample data used with statistical methods. It discusses the use of statistical measures of association, correlation, and regression with relatively uncontrolled and extremely variable natural data.

The book was developed from a symposium of the same name held in Reno, Nevada, in November of 1984 at the Annual Meeting of the Geological Society of America and sponsored by the International Association for Mathematical Geology. The topic of "use and abuse of statistical methods in earch sciences" comes about in part from a lack of communication and understanding between applied statisticians and earth science researchers. Both groups have a common goal in research: to gain a better understanding and description of natural systems. However, the more objective approach of applied statisticians versus the more subjective approach of researchers contributes to a general reluctance by one group to accept the methods of the other. Two gaps in geostatistical methodology exasperate this situation: (1) lack of information on the importance of violations of basic assumptions of probability theory and statistics (e.g., natural data are usually nonrandom, noisy, and nonreproducible) and (2) lack of information on how to translate statistical results into geologically meaningful statements.

Compounding this problem is the ease with which data can be entered into a statistical routine using computers. The effect is to make it more difficult to resolve questions of data reliability, use of appropriate methods, and validity of results. Quality of samples and data, limitations and weaknesses of methods used, and levels of certainty must be considered in order to provide a reliable estimation of the overall soundness of a study. A successful study should combine statistical measures together with the intuition, inference, and experience of the investigator.

Three points of emphasis are made by the authors in this book: (1) an inadequate sampling plan and data base are the cause for many subsequent problems in statistical analysis and inference. Statistical procedures can produce impressive and persuasive-looking results, regardless of quality of data used; (2) there are many mathematical ways to manipulate data statistically, some of which lead to misleading or erroneous results; and (3) the results of statistical analyses should not be taken literally, but rather as approximations (which perhaps can be improved using iterative and feedback methods).

The authors have attempted to discuss applications and interpretations of statistical methods in the earth sciences with minimum support of equations. We begin with a discussion of sample and data reliability and continue with problems in sample mixtures. The theme expands into classification of natural sample data and ends with a discussion on use and abuse of correlation and regression techniques.

The first three chapters deal with the issue of data quality and sample data integrity. The importance of sample definition and representativeness to goals of an investigation must be emphasized, especially since sample data usually are not capable of supporting multiple conclusions. In fact, if the sample size is small and the data variability is great, the precise estimate of statistical measures may be impossible. Natural variance in the sample data should be preserved since it helps put error limits on data to prevent overinterpretation and to point out directions for improving data quality. The degradation of data resulting from spurious samples, transformations, misapplied procedures, and weakly supported interpretations are abundant in the literature.

The next two chapters focus on classification. The objective in the classification of natural objects is to develop a quantitative procedure to recognize natural groups using numerical data from an arbitrary set of observed variables. Although there are certain improprieties in the application of methods such as cluster analysis and discriminant functions using sample data that are nonrandomly and nonuniformly distributed, these procedures can produce a high level of consistency, especially when groupings are widely separated. However, when groups overlap, the credibility or goodness of one boundary line over another is suspect.

The final four chapters deal with regression methods and bivariate correlation. Linear regression, using the least-squares method, is probably the most widely used statistical method in the earth sciences, apart from estimates of mean and variance. A primary purpose of linear regression is to extract the main features of the relationships that are hidden or implied in data. The resulting linear equation can then be used to summarize, infer, predict, compare, or calibrate. Although the linear equation may adequately describe the behavior of the data in a predictive manner, it may be totally useless in a physical or functional manner. Therefore, the purpose for which a regression analysis is to be used should govern selection of variables, not vice versa.

The proper use of the linear regression model is subject to many assumptions and limitations which are frequently, yet unintentionally, violated (see Mann, Chapter 6). For example, the assumption that the independent variable be measured without error makes unrealistic demands on the earth scientist constrained to using highly variable and intercorrelated data. The regression equation may convey a false sense of progress; therefore, it should be only tentatively accepted, while it is critically examined. Accordingly, the investigator should test the adequacy of the linear model using, for example, hypothesis testing and confidence limits.

Translating statistical results of regression into geological inferences of cause and effect requires an understanding of controls in the natural system under investigation. Regression modes are only accurate if the correct predictor population is sampled. In uncontrolled conditions more samples are usually required to explain the different relationships. Extrapolating values for the so-designated dependent variable based on a regression equation can lead to inferences beyond the limits of the data. This is especially true if the designated independent variable has a large error.

Additional problems are encountered in regression analysis when equations of best-fit lines are algebraically manipulated. The equal sign in a regression equation can be very deceptive. The effects of rearranging the equation, or equating two equations that have the same dependent variable, or substitution, multiplication or division of one equation by another are brought to light in Williams and Troutman's chapter. Compound manipulations lead to even more problems, such as combining empirical and theoretical expressions.

Regression models and associated correlation coefficients are only approximations and should not be expected to give exact answers to the inexact study of earth sciences.

Atlanta, Georgia   W.B.S.
September 1986

Acknowledgments

The authors wish to thank the many people who contributed to the writing of this book. Special thanks to the series editors: Richard B. McCammon, C. John Mann, and Thomas A. Jones. The individual authors also wish to thank the following people for their help in preparing their manuscripts: C. W. Hickcox, P. J. W. Gore, P. L. Renwick, R. W. Hayden, E. J. Gilroy, R. H. Meade, G. D. Tasker, J. G. Elliott, and B. W. Chappell.

Contents

Contributors

Theodore J. Bornhorst
Department of Geology and Geological Engineering
Michigan Technological University
Houghton, Michigan

Felix Chayes
Carnegie Institution of Washington
Geophysical Laboratory
Washington, D.C.

Peter Christenson
Department of Biomathematics
University of California at Los Angeles
Los Angeles, California

Robert Ehrlich
Department of Geology
University of South Carolina
Columbia, South Carolina

William E. Full
Department of Geology
Wichita State University
Wichita, Kansas

James A. Harrell
Department of Geology
University of Toledo
Toledo, Ohio

Darrell L. Hicks
Department of Mathematical and Computer Sciences
Michigan Technological University
Houghton, Michigan

Gongshi Li
Department of Geology and Geological Engineering
Michigan Technological University
Houghton, Michigan

C. John Mann
Department of Geology
University of Illinois at Urbana-Champaign
Urbana, Illinois

Paul Schiffelbein
Applied Statistics Group
Engineering Services Division
E. I. DuPont de Nemours and Company
Wilmington, Delaware

William B. Size
Department of Geology
Emory University
Atlanta, Georgia

Brent M. Troutman
Unites States Geological Survey
Water Resources Division
Denver, Colorado

E. H. Timothy Whitten
Vice President for Academic Affairs
Michigan Technological University
Houghton, Michigan

Garnett P. Williams
United States Geological Survey
Denver, Colorado

Use and Abuse
of Statistical Methods
in the Earth Sciences

1

Use of Representative Samples and Sampling Plans in Describing Geologic Variability and Trends

William B. Size

The purpose and representativeness of samples and sampling plans in geologic field-related investigations commonly are given less attention than subsequent analytical and statistical procedures performed on samples. If the appropriateness of a sample to solving a problem is not investigated, the question then becomes: How "good" can results using untested data be? The emphasis is to model the geologic system under study rather than to model numbers obtained from the system.

An investigator's choice of a sample and sampling plan is almost inviolate and usually goes unchallenged. Sampling techniques therefore can become contentious and masked in secrecy. In spite of this apparent weakness, an investigator's experience, skill, and intuition usually provide more geological information in a study than do any structured statistical analyses.

Lack of emphasis on the importance of sample representativeness and sampling plans in geologic field-related problems can be attributed to:

1. *Lack of sample control.* Most preconceived sampling plans cannot be completed because of rather poor control in sample availability and accessibility. Since samples usually must serve several purposes, they are not suited exactly to any one purpose (Williams, 1978).

2. *Lack of normality in data distribution.* Most geologic data are nonuniformly or nonnormally distributed. Such distributions are controlled by

natural processes that develop trends and patterns of much interest to the investigator.

3. *Poor reproducibility of data.* Due to poor control of sample location and the large degree of local variation, geologic data are almost impossible to reproduce and, therefore, reproducibility is usually not attempted.

4. *Inadequately defined objective and model.* In many instances, the initial objective in geologic field-oriented studies is only qualitatively defined using models or results of laboratory experiments (Griffiths and Ondrick, 1968). The true objective and resultant structured model in a study are derived frequently after data are collected.

Factors such as sample accessibility, degree of weathering, cost, and time control the sampled population and the data collected. How "good" or "representative" these survey-type samples are is usually not known without collecting replicated samples to determine levels of variance. For example, whether a hand specimen of basalt taken from an outcrop contains an amount of europium representative of the outcrop is almost impossible to determine. As a result of inadequate sampling in field studies, the population that the collected samples actually represent may or may not relate to the problem and scale of interest of the investigation.

The step from sampled population to target population is usually based on knowledge of subject matter, skill, and intuition, but not on statistical methodology (Whitten, 1966). Any statistical inference that is made applies only to the sampled population. Additional data would be required to decide whether these conclusions are applicable also to the population about which information is actually desired (i.e., subsampling for attributes; Emerson, 1964).

Questions on the appropriateness of a sample or sampling plan to a study are often avoided (intentionally or unintentionally) by overemphasizing analytical precision of instrument-derived data. This frequently gives a false sense of representativeness to the data base and can evolve into varying intensities of conviction or degrees of confidence in the "goodness" of the data. This problem then is complicated further when concise statistical procedures are applied to sample data sets without taking into account relative strengths and weaknesses of the assumptions. Statistical analyses using sample survey data sets should be regarded as merely a useful means to examine data rather than as a technique to provide precise solutions.

SAMPLE DEFINITION AND DESCRIPTION

One of the most qualitative aspects in geologic field-related studies is defining the sample. Commonly, definition and intended purpose of a sample are not fully realized until completion of the analytical stage of the study.

The problem in using poorly defined samples is that they become the subsequent basis for the most exacting analytical procedures, which can generate both accurate and precise machine data. The critical question, however, remains unanswered: Are the data representative of more than just the sample from which they were taken? Are the data representative of their intended population (e.g., the outcrop or perhaps halfway toward the next sampled outcrop)? The target population is the actual object of interest in sampling studies in which generalizations, predictions, and decisions are made. Therefore, samples are of interest only to the extent that they give insight into the target population (Krumbein and Graybill, 1965).

A sample is basically an entity or object about which information or data are sought or collected. Terms used together with a sample such as "representative," "good sense," or "satisfactorily uniform" usually indicate hope. A sample should be representative of the population at the scale of interest and should possess as many of the same characteristics in identical relative degree as do members of the target population.

What constitutes a sample usually becomes a matter of subject-matter experience (although a tendency to sample oddities always exists). Accuracy of such "judgment samples" or "samples of convenience" usually cannot be determined. Judgment samples are not necessarily inaccurate but, if they are, their accuracy is usually unknown and their effectiveness depends on the expertise of the investigator. A statistical sample, however, is chosen by a specified selection method, which includes some degree of randomness. Random samples are required to make decisions and statistical inference on the basis of probability theory.

The representativeness of a sample is critical because the sample acts as a proxy and becomes the basic source of information upon which all subsequent analyses and interpretations will be based. Methods to estimate sample effectiveness involve determination of levels of variance using substage sampling. The literature is relatively sparse on the subject of sample analysis of variance, homogeneity, or representativeness (for a good discussion on the subject, see Koch and Link, 1971; Davis, 1973; or Whitten, 1966).

The scale factor is important in geologic sampling because geologic attributes have varying degrees of homogeneity at different scales (Schryver, 1968). Ideally, an investigator would like to sample at a scale that contains the greatest amount of between-sample attribute variation reflective of processes controlling the value of the attribute at the scale of interest in the study. It is unrealistic to assume that all of the differences in a particular attribute result from controls at the level of interest (e.g., regional trend analysis).

An example of scale factor and homogeneity is given by Kretz (1969) using a detailed drawing of a thin section of granulite containing 1294 grains. He used many tests of distribution and randomness including chi-square, area fraction, point sample, quadrat, nearest neighbor, random point, contact area, and contact frequency analyses. Kretz concluded that the lower limit of homogeneity in mineral proportions in thin section is an area of 30 sq. mm. Any analyzed area less than 30 sq. mm gives unacceptably large variations. Data from Kretz (1969) were used by Fabbri (1984) in a computer-controlled image-analysis study in quantitative petrography. Fabbri's (1984) results on mineral areal proportion and grain boundaries are surprisingly similar to those of Kretz. The potential advantage of rapid, automated model analyses using image-analysis techniques could reduce greatly the effect of bias and nonrepresentativeness of a solitary, point-counted, thin section. Quantitative textural analysis and degrees of textural and structural homogeneity also are possible using image analysis. For example, to estimate textural irregularity, Fabbri (1984) used the complexity index (C.I.), which is the ratio between total circumference and total area of grain profiles for each mineral phase in a thin section.

SAMPLE REQUIREMENTS

Samples should be (1) randomly chosen, (2) representative of the population, (3) sufficiently large to describe the population, and (4) controlled for extraneous variables. Randomness and representativeness are necessary for the proper probability-based statistical reasoning to be made between sample and population. However, because geologic processes control the distribution of natural characteristics, sample data are neither independently nor identically distributed. Sampling and data gathering from field-related geological studies may represent a complex maze in statistical procedure. The pattern of spatial correlation usually is punctuated by such features as faults, lithologic contacts, and alteration zones. The advantage of having samples that are controlled for extraneous variables is that inferences are easier from a homogeneous population. For example, laboratory-derived phase diagrams and models usually are representative of simple systems that best relate, and give inference to, similarly simple field occurrences. The homogeneous sampling unit or analysis unit can be determined by subsampling and analysis-of-variance studies (Ehrlich, 1964; Whitten, 1966; McCall, 1982).

In most field-related problems, the number of samples taken to represent the area under study (i.e., sampled population versus target population) is surprisingly small. For example, to obtain a 1% sample coverage over an area of 1 sq. km would require approximately one million hand samples of rock (10 cm × 10 cm). Therefore, most investigations that involve hundreds or thousands of samples may cover only a fraction of 1% of the population of interest. The more important decision to be made is not whether the number of samples is large enough but, rather, whether the sample is adequate for the accuracy and confidence desired. Accepting the fact that an investigator must work with whatever samples are available, the task then is to (1) try to unravel mixed populations that may have strongly skewed distributions and large variances and (2) identify and minimize multiple sources of bias (Journel, 1984).

SAMPLE VARIANCE AND BIAS

Statistical calculations are blind to any problems or deficiencies in sample data. Yet samples are seldom reanalyzed or subsampled for variation and bias. This avoids confronting the question of what the "true" value of a variable is at the scale of interest. Results of statistical analyses may be shown to be statistically significant, but this does not mean they are necessarily geologically meaningful. For example, LeMaitre (1982) thoroughly investigated the relationship between the depth to the Benioff zone versus potassium percent of lavas (Hutchison, 1975). Questions of randomness, normality, and bias are reflected in the wide range for confidence limits and small correlation coefficient that LeMaitre described between the two variables. Geologic validity of the relationship between increasing potassium content inboard from a convergence zone requires an analysis of variance at the local scale.

Probability theory and its effect on sampling are fundamental to statistical inference. Usually an assumption is made that samples come from some probability distribution. However, because many types of geological data are neither randomly nor normally distributed, care must be taken when applying standard statistical methods to such complex data. The amount of natural variability may also be masked in the collective effects of sample bias, measurement error, and operator error. Chances of introducing these unwanted sources of variation into a field-related study are great because of the many stages involved in obtaining data (e.g., outcrop, hand specimen, thin section, and analyzed powder). For example, the International Study Group (1982) assessed experimental variability in radiocarbon dating among 20 laboratories. They concluded that quoted errors for C-14 dates should be multipled by a factor of 2 to 3.

The judgement of a researcher usually decides what constitutes the minimum acceptable precision or allowable error. Two of the more commonly known sources of error are (1) *statistical error* or precision (reproducibility or random error) and (2) *systematic error,* which relates to sampling variability and refers to the fact that the vagaries of chance will nearly always ensure that one sample will differ from another sample taken by random selection.

Nonsampling errors that can be introduced into field studies include: (1) incomplete coverage, (2) faulty methods of selection, (3) faulty methods of estimation, and (4) enumerator (operator) error. The net effect of these

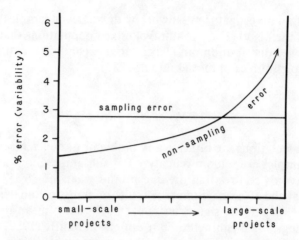

Figure 1.1 The relationship between sampling error (geological variability) and nonsampling error (operator error, machine error, bias) with increasing scale of a project.

contributes to a larger error, rather than tending to cancel out each other. In general, the larger the data set, the larger the percentage of nonsampling error produced (Figure 1.1). The reason larger projects or greater numbers of samples may actually have a greater nonsampling error is that more samples require more operators, more analytical procedures, and so on, which contribute to increasing nonsampling error. Nonsampling errors may be an order of magnitude greater than sampling errors (LeMaitre, 1982).

Any bias has the effect of redirecting a sample estimate of a characteristic away from the true value. The importance of measurement bias is familiar to most investigators; however the importance of selection bias may be overlooked. Selection bias occurs when a sample is not drawn according to a prearranged selection or design. Selection bias is insidious in that all efforts of rechecking measurements and redoing analyses cannot correct the hidden effect of selection bias. Selection bias is the easiest way to mislead or be misled because usually little opportunity exists to check the quality of samples in an investigation.

Natural variability in samples, and the scale at which it is greatest, should be one of the most sought-after unknowns in any quantitative geological study. Variation in a measured characteristic is assumed frequently to be at the scale of interest of the investigator without any inquiry into actual scale of variation. The actual scale of variation may not be the grandiose scale hoped for in an investigation, but it probably will provide a greater understanding of controls and processes affecting its distribution.

Variability and the level at which it is greatest in a sampling study are usually understood in a statistical sense but difficult to translate into a real sense. This uncertainty about the geological use of variance, coupled with a reluctance by investigators to take replicate or subsamples, tends to minimize the importance of locating levels of major variation.

Three ways to describe variation about the estimate of the mean of a measured variable are (1) variance s^2, which, being in squared units, is difficult to comprehend; (2) standard deviation s, which, although in the same units as the variable, relates to a normally distributed population (a rarity in geologic field studies); and (3) "coefficient of variation" C, which is the estimate of standard deviation divided by the mean, $s/\overline{x}$. Other names for the coefficient of variation are "relative error" and "relative standard deviation."

Although the coefficient of variation is not used frequently in statistical analysis, it is readily calculated and can be more easily interpreted compared to other measures of distribution. It gives a clearer description of the spread, because it takes into account both mean and variance. Multiplying the coefficient of variation by 100 expresses standard deviation as a percentage of mean value. An example of the use of the coefficient of variation is given by Koch and Link (1971). The relationship between coefficient of variation and geochemical percentage data (Figure 1.2, stippled zone) shows that the coefficient of variation increases with decreasing amount of a chemical component in the sample. In parts per million range, variation

Figure 1.2 Relationship between the coefficient of variation and average values from percentage data. The coefficient of variation increases as the concentration of the component decreases. The horizontal line at $C = 0.5$ represents the boundary between normally distributed data and nonnormally distributed data (Koch and Link, 1971).

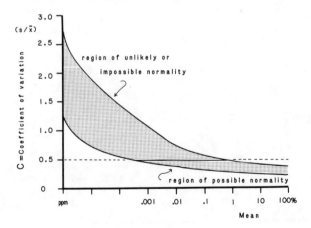

can be 100 to 300% of the mean. The horizontal line drawn at a coefficient of variation of 0.5 (50%) marks the boundary between data that are unlikely to be normally distributed (greater than 0.5), from data that are likely to be normally distributed (less than 0.5; Koch and Link, 1971).

The coefficient of variation can be used by itself to get a feeling for variability at any scale desired or it can be used together with an analysis-of-variance table. Table 1.1A shows an example of chemical and mineralogical data from replicated samples of homogeneous trondhjemite (Size, 1979, 1984).

All samples came from freshly exposed rock in a small quarry in central Norway. Three hand specimens of trondhjemite taken one meter apart in a triangular pattern comprised each sample. From each hand specimen, two thin sections were analyzed using the point-counting method of Chayes (1956; 1200 counts per slide), with the remainder of the hand specimen powdered for chemical analysis. Analysis-of-variance and coefficient of variation results for modal quartz and whole-rock SiO_2 weight % (Table 1.1B) indicate that coefficient of variation for modal quartz ranges from 3.2 to 15.4% of the mean. On a somewhat heuristic basis, coefficients of variation greater than 10% of the mean warrant closer inspection before an estimate of the mean is used as a proxy for the population mean. Only one sample (Table 1.1B) has a coefficient of variation for modal quartz greater than 10% (TR-43). The average of 30 thin-section analyses of trondhjemite shows a coefficient of variation of 8.4%, which indicates an acceptable spread about the mean (i.e., modal quartz content is rather

TABLE 1.1A Means and Standard Deviations for Whole-Rock Geochemistry and Modal Mineral Percentages from the Type-Area Trondhjemite Central Norwegian Caledonides (Follstad, Norway)

Wt. %	$\bar{x}$	s	Vol. %	$\bar{x}$	s
SiO_2	71.47	(0.48)	Plagioclase	57.8	(4.3)
TiO_2	0.23	(0.02)	K-feldspar	3.3	(2.9)
Al_2O_3	16.20	(0.34)	Quartz	26.1	(2.2)
Fe_2O_3	0.58	(0.08)	Muscovite	5.9	(2.0)
FeO	0.78	(0.04)	Biotite	1.9	(1.0)
MnO	0.02		Epidote	4.4	(1.1)
MgO	0.48	(0.05)	Sphene	0.2	
CaO	2.72	(0.33)	Magnetite	tr.	
Na_2O	5.33	(0.14)	Apatite	tr.	
K_2O	1.39	(0.15)	Chlorite	tr.	
P_2O_5	0.08	(0.01)	Calcite	tr.	
H_2O	0.58	(0.16)			

$n = 15$ $n = 30$

Source: Size (1979).

TABLE 1.1B Analysis-of-Variance Tables for Modal Quartz Percentages and Whole-Rock SiO_2 Percentages from the Same Replicated Samples of Trondhjemite Shown as Means in Table 1.1A

Analysis of variance of percentage of modal quartz (1200 counts per thin section) from replicated samples of trondhjemite (data from Size, 1979).

Rock sample number	Number of thin-section replicates	Mean quartz content	Standard deviation	Coefficient of variation
TR-43	6	25.5%	3.9%	15.4%
TR-45	6	27.2%	1.3%	4.9%
TR-46	6	26.0%	0.8%	3.2%
TR-47	6	25.3%	1.3%	5.3%
TR-49	6	26.3%	2.2%	8.5%
Total	30	26.1%	2.2%	8.4%

Analysis of variance table for modal quartz between-sample variation versus within-sample Variation.

Source of variation	Degrees of freedom	Sum of squares	Mean square	F-value[a]
Between samples	4	13.41	3.33	0.68
Within sample	25	123.02	4.92	
Total	29	136.43		

[a]Critical value of $F_{4,25} = 4.18$ for alpha 0.01.

Analysis of variance of whole-rock SiO_2 weight percentages from replicated samples of trondhjemite (data from Size, 1979).

Rock sample number	Number of rock powder replicates	Mean SiO_2 wt. % (whole rock)	Standard deviation s	Coefficient of variation
TR-43	3	71.58%	0.14%	0.20%
TR-45	3	71.54%	0.25%	0.35%
TR-46	3	71.24%	0.26%	0.36%
TR-47	3	70.84%	0.09%	0.12%
TR-49	3	72.14%	0.33%	0.46%
Total	15	71.47%	0.48%	0.67%

TABLE 1.1B (*Continued*)

Analysis of variance table for whole-rock SiO_2 weight percentages between-sample variation versus within-sample variation

Source of variation	Degrees of freedom	Sum of squares	Mean square	F value[a]
Between samples	4	2.72	0.68	13.53
Within sample	10	0.54	0.05	
Total	14	3.26		

[a]Critical value of $F_{4,10} = 5.99$ for alpha 0.01.

homogeneous). This is supported in the analysis-of-variance table, which compares the between-sample variance in modal quartz with within-sample variance (Table 1.1B). The F value at 4 and 25 degrees of freedom is only 0.68, well below the F-table value of 4.18 at identical degrees of freedom with an alpha value of 0.01. Results indicate that no statistical evidence suggests that the samples came from different populations.

The variance description for SiO_2 whole-rock weight percent (Table 1.1B) displays a misleading F value of 13.53. This is more than the critical F value of 5.99 for 4 and 10 degrees of freedom with an alpha level of 0.01. The null hypothesis is therefore rejected, implying that some real differences between samples may be present. However, a closer look at the data and coefficient of variation indicates a different conclusion. The sample mean SiO_2 weight percentages are close to the overall mean value and all of the coefficients of variation are less than 1%. The reason lies in the extremely small value for total sum of squares (3.26). Slight differences in within-sample versus between-sample source of variation would change the F value considerably. In this case, the coefficient of variation provides a more realistic appraisal of homogeneity of data than the F- value.

SAMPLING PLANS

The relative merit of one sampling plan over another is usually a moot question in field-oriented geologic problems because the availability of samples is so well controlled by outside influences. The plan becomes one of "feasible sampling" or "available sampling," which is usually nonrepresentative and not based on any probability distribution. This type of sampling is composed of "typical" samples selected by the investigator after some inspection. This method of controlled sampling is also called "judgment selection" or "purposive selection" (Cochran, 1963).

Sampling is the selection of part of an aggregate to represent the whole in order to obtain information about a population. The goal of sampling is to find a sampling scheme that fits the specific characteristics of the target population (Olea, 1984). Efficiency of a sampling design depends greatly on knowledge of the magnitude of components of variance contributed at different levels (i.e., by performing a preliminary or survey sampling). The real impetus should be to discover at what level the greatest amount of variance occurs for each variable of interest and then to describe this variation as completely as possible (McBratne, Webster, and Burgess, 1981).

As a result of an inadequate sampling plan (i.e., nonrandom and nonnormal distribution and nonrepresentative), a population is introduced into the study which is the population that samples actually represent. This sampled population may be typical for certain properties, but totally inappropriate for less easily observed characteristics. The desire to describe these properties makes conflicting demands on a sampling design, which usually cannot be accommodated without a great increase in the number of individuals and samples.

Although many sampling projects begin by deciding how many samples will be taken for the study, sample size determination should be postponed until levels of variance have been estimated for the variables of interest. The emphasis in sampling should not be placed on the number of samples taken from a population, but rather on the number of individuals in the sample. The greater the level of heterogeneity, the more individuals are necessary to describe the variation. Iterative sampling and analysis of variance can provide the necessary advance information about the population and its characteristics in order to design a more effective sampling plan.

Increasing sample size has a desirable effect of reducing sample variance, but this reduction decreases progressively. In order to cut sampling variance in half, sample size must be doubled. Therefore, the biggest gains in efficiency can be made using clever sampling designs and estimators, rather than increasing sample size. If no statement of precision is to be made, any sample size will do!

In spite of a lack of control of sample locations and sampling inadequacies, once an outcrop, roadcut, or drill hole has been chosen, an opportunity exists to determine local variation in characteristics of interests by using sublevel random selection. This is a necessary step in determining how representative an estimate of a variable is to its intended purpose.

Examples of sampling plans

Sampling plans should include components of randomness, coverage, and representativeness (Sukhatme et al., 1984). Statisticians tend to use a great number of random samples in their studies, thus avoiding the question of

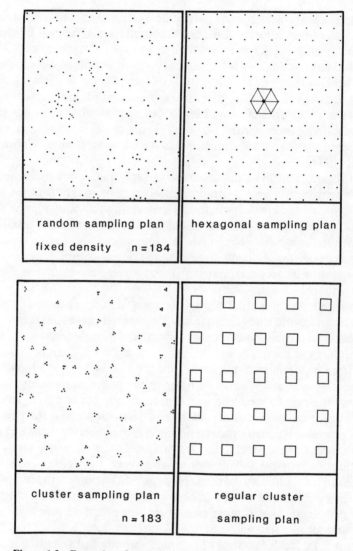

Figure 1.3 Examples of random, ordered, and cluster sampling plans with constant number of samples.

representativeness (i.e., a great number of random samples probably is representative also). Random sampling plans can, at best, only be approximated in geologic field-related studies (Figure 1.3). This is not just due to lack of sample availability at random locations, but is also due to the justified reluctance of investigators to forfeit all control of sample location to probability theory. In addition, some question by investigators exists about return compared to effort expended for usable geological information from random sampling plans.

Another concern investigators have is that random sampling of points tends to leave holes or gaps in sample distribution. Natural processes tend to produce trends, clusters, and punctuations in data distribution that can be missed in random sampling. Therefore, *coverage* becomes relatively more important than randomness in the sampling plan. The hexagonal-net sampling plan (Figure 1.3) has the same number of points as the random sampling plan ($n = 184$). The points are all equidistant from one another, producing a uniform coverage. However, because of the lack of randomness in this plan, no probability-based statistical inference should be made, although the investigator can still make geological inferences.

Two examples of cluster sampling plans are also shown in Figure 1.3: a random cluster distribution and a regular cluster distribution. Size of clusters will influence efficiency of samples. The smaller the cluster, the more accurate will be the estimate of sample characteristics. The optimum cluster is one that estimates the characteristics under study with maximum precision for a given proportion of the population studied. The procedure of first selecting a cluster (e.g., outcrop) and then choosing a specified number of elements (e.g., hand specimens) from the cluster at random is known as subsampling or multistage sampling.

Cluster sampling is recommended for geologic field-related studies for three reasons: (1) the investigator regains control of the sample location, (2) a certain degree of randomness can be introduced in the location of individuals within a cluster, and (3) multistage sampling allows for analysis of variance (ANOVA).

Stone Mountain, Panola Mountain, and Mount Arabia represent three granitic rock bodies in the Georgia Piedmont (Grant et al., 1980). They are near one another and of similar age (ca. 325 MA). Cluster sampling from each body included (1) nonrandomly choosing an outcrop, (2) picking two hand specimens at random, (3) cutting two thin sections at random from each hand specimen, and (4) performing two (1200 count) modal analyses on each thin section by turning the slide around. In retrospect, at least three individuals should have been collected at each level to provide a better basis for estimating mean and variance.

Means and standard deviations for modal quartz were determined (Table 1.2) along with ANOVA tables for three levels of comparison: (1) variation within rock samples from each outcrop, (2) variation between rock samples from each outcrop, and (3) variation within thin sections between outcrops. Quartz variation between outcrops (last column, Table 1.2) for both Stone Mountain and Panola Mountain have F values greater than the tabled value for 1 and 14 degrees of freedom ($F = 4.60$). This indicates that some real differences in quartz content between outcrops may exist. However, evidence exists that local (within outcrop) variation is considerable. Sample 3 from Stone Mountain has an F value of 183.6 for within rock sample variation and outcrop 1 from Panola Mountain has a value of 155.52 for between rock sample variation from each outcrop.

TABLE 1.2 Analysis-of-Variance Table for Modal Quartz Percentages from Replicated Samples of Granitic Rocks from Three Nearby Crystalline Bodies in the Central Georgia Piedmont

Analysis of variance of quartz modal mineral percentages on three levels		Variation within rock samples from each outcrop		Variation between rock samples from each outcrop		Variation within thin sections between outcrops	
		Mean and standard deviation (2 modes each)	$F_{.05}(1,2) = 18.5$ (F-value calculation)	Mean and standard deviation (4 modes each)	$F_{.05}(1,6) = 5.99$ (F-value calculation)	Mean and standard deviation (8 modes each)	$F_{.05}(1,14) = 4.60$ (F-value calculation)
Stone Mountain							
Outcrop 1 Sample 1	Slide A	38.35(1.90)	0.39	39.9(4.4)	0.01	39.66(5.0)	5.23
	Slide B	41.45(6.70)					
Sample 2	Slide A	43.75(4.30)	3.36	39.42(6.3)			
	Slide B	35.10(5.00)					
Outcrop 2 Sample 3	Slide A	39.70(0.71)	183.6	35.75(4.6)	0.70	34.67(3.5)	
	Slide B	31.80(0.42)					
Sample 4	Slide A	35.15(0.78)	3.18	33.6(2.3)			
	Slide B	32.05(2.30)					
Panola Mountain							
Outcrop 1 Sample 1	Slide A	21.70(0.99)	1.09	21.32(.73)	155.52	25.53(4.59)	7.56
	Slide B	20.95(0.21)					
Sample 2	Slide A	29.80(1.80)	0.005	29.75(1.1)			
	Slide B	29.70(0.71)					
Outcrop 2 Sample 1	Slide A	29.80(1.13)	0.031	29.87(.69)	0.47	30.05(0.68)	
	Slide B	29.95(0.35)					
Sample 2	Slide A	30.70(0.71)	2.42	30.22(.74)			
	Slide B	29.75(0.49)					
Mt. Arabia							
Outcrop 1 Sample 2	Slide A	37.65(1.48)	4.70	Granite 31.15(4.82)	0.17	34.67(3.92)	0.004
Granite	Slide B	30.65(4.31)		Gneiss 35.7(1.98)			
Outcrop 2 Sample 3	Slide A	31.30(0.71)	8.40	Granite 35.92(5.9)	0.59	34.78(4.94)	
Granite	Slide B	40.55(4.40)		Gneiss 32.5(.56)			

Source: Grant et al. (1980).

16

The last example at Mt. Arabia shows a small between-outcrop F value (0.004), indicating some degree of homogeneity. However, the within-outcrop coefficient of variation is 11.3% for outcrop 1 and 14.2% for outcrop 2, indicating that an important source of within-outcrop variation may exist that warrants further investigation. Such levels of local variation, important in their own right, should also caution an investigator about further application of the data in regional trend analysis.

Sampling plans and trends

Earth processes control distribution of many characteristics of interest to geologists who are trained to detect such trends and patterns in nature (e.g., maps, structures, petrography, fossils). The desire to discover trends in geologic data is specially keen at the regional scale. In fact, the failure of an investigator to discover patterns or trends in data frequently reflects on the quality of research.

Nature ensures that some degree of heterogeneity always will be present in a variable measured from samples. However, to assume that variability is most important at the same level of interest to an investigator is unsound. Sampling plans intended to detect trends must include subsampling for local variation and attributes. It is impossible to appreciate the validity of any trend if spatial distribution and variability of sample data are not given (Whitten, 1961; Mathews et al., 1975; Peikert, 1965).

Two points concerning geologic trend analysis are worth emphasizing. First, most field-related studies are actually no more than descriptive surveys and should only be used to outline the general characteristics of the population (i.e., straw poll). Second, trend surface analysis uses numbers as input, with no consideration to geological circumstances. Therefore, analysis depicts only trends in numbers, whereas the investigator is interested in trends of geologic characteristics. The important difference between these two types of trends is that results of a trend analysis may actually be geologically correct but no way of measuring this exists unless the sampling plan included subgroup variability.

In trend analysis two assumptions are made: (1) the sample value is expected to range from place to place and (2) there is expected to be random variation composed of (a) a "fixed" component of the trend (which is probably unknown) and (b) a random variation following a specific distribution. Some mathematical function usually relates differences between samples, but that function may not be operating at the scale of interest. The function may actually be operating at a scale that is much more informative.

Statistical testing of trends involves an analysis of variance and F distribution. ANOVA compares the source of variation due to regression (trend) with the source of variation due to deviation (unaccounted for variation) (Davis, 1973). Calculated deviations are from the mathematical procedure and not necessarily from natural local variation (Krumbein, 1972). Any measured variable whose between site variability (regional trend) is not significantly greater than within site variability (local, natural variation) will not produce meaningful trend surfaces.

Pitfalls of using ANOVA and F distributions to test the significance of trends are hidden in the dazzle of attractive trend surface and hypersurface maps having impressive correlation coefficients assigned to them. If the original data are not representative or if they have a large, local, natural variation, the geological results of any trend-surface analysis using this data are pure speculation.

RECOMMENDATIONS AND CONCLUSIONS

The following list contains, for the most part, intuitively obvious statements about a commonsense approach to most any type of geological field investigation involving statistical analysis of a relatively small number of widely spaced samples.

1. An investigator should always be involved in sampling design, sample analysis, data analysis, statistical analysis, and inference stages of a study. This person knows the representativeness of samples and data better than anyone else and therefore is best suited to pick out spurious or unreliable features.

2. Limitations of samples and sampling plan for subsequent analytical and statistical analyses should be fully described and explicitly stated.

3. Do not underestimate the effect of bias in an investigation. This contribution to variance should be identified and reduced as much as possible. Selection bias requires special attention in regard to representativeness.

4. Stay as close to the original data as possible. Statistical analysis using second and third generation data, such as ratios, can be misleading (i.e., develop their own patterns; see Mann, Figure 6.3, Chapter 6).

5. Do not expect great revelations from any statistical procedure. Chances are that if a pattern or trend is not obvious before performing a statistical analysis, it is probably so weak that its real existence is questionable (even though it may still be shown to be statistically significant). The purpose of geostatistics is not to investigate or model numbers but to investigate and model the geologic system under study.

6. Complexity in the sources of natural variance should be explored fully and described. Discovery of contributions to variance at different lev-

els could well provide a wealth of information not fully appreciated before.

7. The coefficient of variation, $C = s/\bar{x}$ can provide a clear picture of the spread of data around an estimated mean. Local variation (i.e., within outcrop or within sample variation) with a coefficient of variation greater than 10% of the mean indicates that variation in the measured constituent at that level warrants closer inspection for geologically meaningful patterns.

8. Experiment with the weighting control in the data matrix by selectively removing data and then rerunning the statistical analysis. Major changes in results require a closer look at the reliability of the data removed from the matrix.

9. Be cautious of ANOVA results for testing of significance of trends. The coefficient of variation for local, natural variation should be included in the test. Local, natural variation should at least be less than the trend-surface contour interval.

10. Include some type of multistage or cluster sampling plan using randomly chosen individuals at the local scale. This type of sampling plan, together with ANOVA and coefficient of variation, will aid in locating geologically important sources of variability in characteristics of interest.

REFERENCES

Chayes, F., 1956, *Petrographic modal analysis; An elementary statistical appraisal:* Wiley, New York, 113 p.

Cochran, W. G., 1963, *Sampling techniques:* Wiley, New York, 413 p.

Davis, J. C., 1973, *Statistics and data analysis in geology:* Wiley, New York, 550 p.

Ehrlich, R., 1964, The role of the homogeneous unit in sampling plans for sediments: *Jour. Sed. Pet.,* v. 34, p. 437–439.

Emerson, D. O., 1964, Modal variations within granitic outcrops: *The American Mineralogist,* v. 49, p. 1224–1233.

Fabbri, A. G., 1984, *Image Processing of Geological Data:* Van Nostrand Reinhold, New York, 244 p.

Grant, W. G., W. B. Size, and B. J. O'Connor, 1980, Petrology and structure of the Stone Mountain granite and Mount Arabia migmatite, Lithonia, Georgia, *in* Frey, R. W., and T. L. Neathery, (eds.), *Excursions in southeastern geology:* American Geological Institute, Falls Church, Va., p. 41–57.

Griffiths, J. C., and C. W. Ondrick, 1968, Sampling a geological population, *in* Merriam, D. F. (ed.), *Computer Contribution 30:* Kansas State Geological Survey, University of Kansas, Lawrence, Kans., p. 1–53.

Hutchison, C. S., 1975, Correlation of Indonesian active volcanic geochemistry with Benioff zone depth.: *Geol. Mijnbouw,* v. 54, p. 157–168.

International-Study-Group, 1982, An inter-laboratory comparison of radiocarbon measurements in tree rings: *Nature,* v. 298, p. 619–623.

Journel, A. G., 1984, Geostatistics, simple tools applied to difficult problems, *in* David, H. A., and H. T. David (eds.), *Statistics: An appraisal:* The Iowa State University Press, Ames, Iowa, p. 237–256.

Koch, G. S., Jr., and R. F. Link, 1971, *Statistical analysis of geological data:* Wiley, New York, 375 p.

Kretz, R., 1969, On the spatial distribution of crystals in rocks: *Lithos,* v. 2, p. 39–66.

Krumbein, W. C., 1972, Areal variation and statistical correlation, *in* Merriam, D. F. (ed.), *Mathematical models of sedimentary processes:* Plenum Press, New York, p. 167–174.

Krumbein, W. C., and F. A. Graybill, 1965, *An introduction to statistical methods in geology:* McGraw-Hill, New York, 475 p.

LeMaitre, R. W., 1982, *Numerical petrology:* Elsevier, Amsterdam, 281 p.

Mathews, G. W., J. A. Cain, and P. O. Banks, 1975, Three-dimensional polynomial trend-analysis applied to igneous petrogenesis, *in* Whitten, E. H. T., (ed.), *Quantitative studies in the geological sciences:* The Geological Society of America, Boulder, Colo., p. 239–256.

McBratney, A. B., R. Webster, and T. M. Burgess, 1981, The design of optimal sampling schemes for local estimation and mapping of regionalized variables—I, theory and method: *Computers and Geosciences,* v. 7, p. 331–334.

McCall, C. H., Jr., 1982, *Sampling and statistics; handbook for research:* The Iowa State Univ. Press, Ames, 340 p.

Olea, R. A., 1984, Sampling design optimization for spatial functions: *Math. Geology,* v. 16, pp. 369–392.

Peikert, E. W., 1965, Models for three-dimensional mineralogic variation in granitic plutons based on the Glen Alpine stock, Sierra Nevada, California: *Bull. Geol. Soc. Amer. 76,* p. 331–348.

Schryver, K., 1968, Precision and components of variance in the modal analysis of a coarse-grained augen gneiss: *The American Mineralogist,* v. 53, p. 2036–2046.

Size, W. B., 1979, Petrology, geochemistry and genesis of the type area trondhjemite in the Trondheim region, central Norwegian Caledonides: *Norges Geologiske Undersokelse,* v. 351, v. 51–76.

Size, W. B., 1984, Polygenetic trondhjemite, *in Proceedings of the 27th international geological congress,* Moscow, VNU Science Press, Utrecht, the Neatherlands p. 543–559.

Sukhatme, P. V., B. V. Sukhatme, S. Sukhatme, and C. Asok, 1984, *Sampling theory of surveys with applications:* Iowa State University Press, Ames, 526 p.

Whitten, E. H. T., 1961, Quantitative areal modal analysis of granitic complexes: *Geol. Soc. Amer. Bull.* v. 72, p. 1331–1360.

Whitten, E. H. T., 1966, *Structural geology of folded rocks:* Rand McNally, Chicago, 663 p.

2

Calculation of Confidence Limits for Geologic Measurements

Paul Schiffelbein

Geologic data are used in a variety of ways. Data are used in their raw form, or they are processed in some way (e.g., smoothing, deconvolution, factor analysis, and regression). In all cases, however, the data are interpreted to extract some information about a geologic process. Proper data interpretation can only be accomplished when confidence limits are known for the data. What is the significance of a small "wiggle" in a stable isotope stratigraphic record or an apparently bimodal grain size distribution? Confidence limits protect against overinterpretation of data and point out directions for improving data quality. In this chapter we will examine some aspects of data quality—establishment of precision and confidence measures—and applications of those sampling theory principles to geologic data.

PRECISION AND CONFIDENCE LIMITS

Quantities that can be computed from population data, such as population mean or variance, are generally referred to as population parameters. Similar quantities computed from sample data are referred to as sample statistics. If our observations or measurements can be considered random samples from the population of interest, it is reasonable to use the sample data as a basis for conclusions about the population. That is statistical inference.

Geological measurements are often made to obtain estimates of the magnitude of population parameters on the basis of sample statistics. That type of inference is referred to as point estimation. A point estimate will almost never be correct in the sense of equaling the parameter. Additionally, the extent of error of the estimate cannot be judged on the basis of the estimate itself, since the estimate does not contain information about its sampling distribution. In order to evaluate the adequacy of a point estimate we need to have an estimate of certain limits within which the parameter falls, and a quantitative statement of confidence that the parameter falls within those limits. We will concern ourselves with establishing confidence limits on the population mean μ.

The determination of a confidence interval for μ requires the following information:

a. Value of the sample mean.

b. Value of the population standard deviation or some estimate of it.

c. Equation for the population frequency distribution.

d. Size of the sample.

e. Degree of confidence required.

That last item is decided upon by the experimenter and will always be a number between zero and one.

If our measurements are normally distributed, we can write the probability statement (Bendat and Piersol, 1971):

$$\text{Prob}\left[\hat{\mu} - \frac{\hat{\sigma} t_{n;\alpha/2}}{\sqrt{N}} \leq \mu < \hat{\mu} + \frac{\hat{\sigma} t_{n;\alpha/2}}{\sqrt{N}} \right] = 1 - \alpha$$

which is read "the probability that the population mean, μ, is between $\pm \frac{\hat{\sigma} t_{n;\,\alpha/2}}{\sqrt{N}}$ is $1 - \alpha$," where N is the number of individuals in the sample, $n = N-1$ are the degrees of freedom, $1-\alpha$ is the confidence coefficient associated with the interval, and $\hat{\mu}$ is the calculated sample mean. The confidence coefficient is usually expressed as a percentage. The above equation means that precision of the mean value estimate is less than or equal to one-half the length of the confidence interval at a confidence level of $(1 - \alpha)\%$. (Yamane, 1967). Three variables affect the width of the confidence interval and, hence, the precision of the estimate: $\hat{\sigma}$, N, and the confidence level.

DATA VARIABILITY AND SAMPLING FRAME

Variability of geologic data can be determined in a number of ways. Intrasample stable isotope data variability ($\delta^{18}O$ and $\delta^{13}C$) has been examined in deep-sea sediments using multiple analyses of single foraminifer tests or small numbers of tests from a single sediment sample (Killingley et al., 1981; Schiffelbein and Hills, 1984). Variability thus determined is considerably larger than mass spectrometer measurement error, indicating a large amount of test-to-test isotopic variability in the population. That sampling approach is unsatisfactory for estimating variability of a large number of geologic variables, however. Species counts, % foraminifer fragment counts (often used as a dissolution or preservation index), and grain size measurements are made after a sample has been washed, sieved, or otherwise homogenized. Repeated analyses will bring out only the measurement precision, which will likely be small compared with true variability of those parameters in nature. Estimates of stable isotope data variability will also be optimistically small when based on repeated analyses from a single homogenized sample.

In simple random sampling, variance of the estimate depends both on the sample size and variability of the character in the population. If the population is extremely heterogeneous (most natural populations are) and sample size is limited, a sufficiently precise estimate of the variability may be impossible to obtain.

A sampling frame is defined as materials or devices that delimit, identify, or allow access to elements of the target population. (Wright and Tsao, 1983, provide an excellent summary of the concept of frame.) In our case, the frame encompasses a number of items, including size of samples, number of samples, measurement techniques, and methods of estimation. The quality of an estimate of population variance depends largely upon the condition and adequacy of the frame from which the sample is selected. A study focused on measurement sampling error, which, for argument's sake, may quantify intrasample variability, will see large intersample variability as nonsampling error. Extent of nonsampling error may far exceed that of sampling error. Much of that so-called nonsampling error can be attributed to a poor frame.

The important aspect of frame here is "the population of interest." We wish to determine variability about the mean value of certain geologic variables in a meaningful way, that is, variability in nature. If stratigraphic measurements show lateral heterogeneity, an attempt should be made to include that variability in the estimate. An estimate of variability based on repeated analyses of a single sample would, in that case, lead to erroneous results because of deviation of the target population from the sampled population (i.e., the population that matches the frame). Most estimates of population variability in geological systems would likely be of that sort and would lead to an underestimation of true variability.

ESTIMATING VARIABILITY

When dealing with real populations, unjustified assumptions concerning the underlying distribution should be avoided. In particular, classical tests and confidence intervals for variance are extremely sensitive to deviation of the sample distribution from normality (Box, 1953; Miller, 1968; Mosteller and Tukey, 1977). As a consequence, an increasing interest exists in the development of robust statistical methods that do not require an assumption of distribution normality, but still retain some desirable attributes commonly associated with tests based on that assumption. One such technique, the jackknife, has proved to be a robust and powerful statistical tool.

The jackknife is used to estimate confidence limits for an unknown parameter whose population distribution cannot be assumed to be normal. (Miller, 1974 and Efron, 1982 provide excellent reviews of jackknife theory and procedure.) In the general case, the jackknife is an example of a resampling scheme, where sampling properties of a statistic are examined by computing its value for various subsamples of the original data set. Let θ be an unknown parameter, and let $X_1, X_2, \ldots, X_N$ be N independent identically distributed observations from the distribution function F_θ. The essence of the jackknife is to divide the N observations into p groups of k observations each ($N = pk$); that is, $(X_1, \ldots, X_k), (X_{k+1}, \ldots, X_{2k}), \ldots, (X_{(p-1)(k+1)}, \ldots, X_{pk})$. That division may be determined by the structure of the experiment or may be arbitrarily imposed. Let $\hat{\theta}$ be the estimator of θ based on all N observations and let $\hat{\theta}_i, i = 1, 2, \ldots, p$, denote the estimator of θ defined on the subsample obtained after deletion of the ith group of observations. We define

$$J_i(\hat{\theta}) = p\hat{\theta} - (p - 1)\hat{\theta}_i, \qquad i = 1, 2, \cdots, p$$

and

$$J(\hat{\theta} F\theta) = \frac{1}{p} \sum_{i=1}^{p} J_i(\hat{\theta}) = p\hat{\theta} - \frac{p-1}{p} \sum_{i=1}^{p} \hat{\theta}_i$$

Estimators $J_i(\hat{\theta})$ are called "pseudovalues" of the jackknife and the average of those values, $J(\hat{\theta})$, is called the jackknife (Gray and Schucany, 1972).

Quenouille (1956) originally conceived the jackknife to achieve a reduction in estimator bias. Tukey (1958) later proposed that in most instances, $J_i(\hat{\theta})$ values can be treated as p approximately independent identically distributed observations. The validity of that suggestion, which allows the jackknife to be used for confidence interval construction, has subsequently been established for a large number of cases.

Variability of geological measurements is examined in this chapter using multiple samples from a single stratigraphic horizon. That variability is a combination of measurement error (machine or operator error plus error from finite sampling) and variability of the measured property in nature. Variability thus determined includes, at some scale, effects of sediment heterogeneity within a given stratum, and gives a truer indication of data reproducibility in most cases than does variability determined from repeatedly subsampling a single sample.

Data for this study were collected from equatorial Pacific piston core ERDC 84P (1° 25.3'N; 157° 15.3'E; 2339 m water depth; sedimentation rate 1.6 cm/ka). The 9 cm diameter piston core was first split lengthwise into "archive" and "working" halves. A segment of the "working" half of the core was split lengthwise into four strips each approximately 2 cm wide, and continuous 2 cm thick samples were taken along the core. That yielded four separate samples at each stratigraphic level.

The data consist of 21 groups of 4 subsample measurements, or 84 measurements of each variable. (Data are listed in Appendix I of Schiffelbein, 1984). Those data cannot be treated in a conventional way because each group is from a different stratigraphic level, that is, mean values change from group to group. We are interested in between-sample variance at each stratigraphic horizon, but not variance between groups of measurements at different stratigraphic levels.

The ability of the jackknife to operate on grouped observations gives us a straightforward method for examining data variability within strata, while avoiding the effects of stratigraphic changes in the measurements (process changes). A variance was first calculated for each group of four measurements. The estimator we are interested in is the square root of the sample variance, $\hat{\sigma}$, and was formed by averaging the 21 group variances and taking the square root. Estimators $\hat{\sigma}_i$ were formed similarly and used to calculate the jackknife pseudovalues. Jackknife estimates of population standard deviation were calculated as in Gray and Schucany (1972, p. 166–167). Because we are interested in an upper bound on variance at a given confidence level, we calculate an upper one-sided confidence interval. That is given by

$$J(\hat{\sigma}) + t_{n;\alpha} \sqrt{\operatorname{var} J(\hat{\sigma})}$$

where $J(\hat{\sigma})$ is the jackknife estimate of the population standard deviation, $t_{n;\alpha}$ is the one-tailed t-variate (1-α point) with n degrees of freedom, and $\sqrt{\operatorname{var} J(\hat{\sigma})}$ is the estimated standard deviation of the jackknife estimate, that is,

$$\sqrt{\operatorname{var} J(\hat{\sigma})} = \left\{ \frac{1}{p-1} \sum_{i=1}^{p} [J_i(\hat{\theta}) - J(\hat{\theta})]^2 \right\}^{1/2}$$

Jackknife estimates of population standard deviation and confidence limits for a variety of data are given in Table 2.1. Wt. % > 63 μm was determined by weighing the dry bulk sample, washing the sample over a 63 μm sieve, and weighing the residual. The larger grain size interval was determined by dry sieving. *Pulleniatina* fragments were counted in the 350–420 μm size fraction. Fragments and whole individuals were counted

TABLE 2.1 Jackknife Estimates of Population Standard Deviations

Variable	Degrees of freedom	J $(\hat{\sigma})$	$\sqrt{\text{var J} (\hat{\sigma})}$	80%	90%	95%	97.5%	99%
Wt. % > 63 μm (II)	24	2.67	1.04					
Wt. % > 63 μm (V)	23	2.58	1.19					
Wt. % > 63 μm (II + V)	48	2.62	1.08	3.54	4.02	4.43	4.79	5.22
Wt. % 350–420 μm	18	0.48	0.20	0.65	0.75	0.83	0.90	0.99
% *Pull.* fragments	24	5.63	2.45	7.73	8.86	9.82	10.69	11.71
% *G. sacculifer*	24	2.78	1.41	3.99	4.64	5.19	5.69	6.29
% *N. dutertrei*	24	2.25	1.03	3.13	3.61	4.01	4.38	4.82

until 100 whole specimens were found (average whole specimens + fragments counted = 140). Two foraminifera species (*Globigerinoides sacculifer* and *Neogloboquadrina dutertrei*) were counted in the 350–420 μm size fraction. At least 300 individual foraminifers were counted in each sample (the average number of foraminifers counted was 330). All of the data are from an interval spanning Termination II (5/6 oxygen isotope stage boundary: $\sim$125 kaBP) with the exception of one set of grain size measurements from Termination V (11/12 oxygen isotope stage boundary: $\sim$ 460 kaBP). Those are marked in the table. The two wt. % > 63 μm data sets showed similar variability and were combined for the confidence interval construction.

Stable oxygen and carbon isotope data were also collected from the subsamples. Those data require a slightly different treatment. Following Killingley et al. (1981), two sources of error variance are identified in the measurements: machine noise, σ_m^2, and intrasample variability, σ_s^2. Those sources are assumed independent and, therefore, additive. If total measurement variance is denoted by σ_t^2, then

$$\sigma_t^2 = \sigma_m^2 + \sigma_s^2$$

The intrasample variability term, σ_s^2, is variance of the distribution of sample means, $\hat{\mu}$, about the population mean, μ. Precision of the estimator for a sample mean then is defined as $\tilde{\mu} = \mu - \hat{\mu}$. Denoting the observations by x and variance of the population from which they are drawn by σ^2, then

$$\hat{\mu} = 1/N \sum_{i=1}^{N} x_i$$

and

$$\text{var} (\tilde{\mu}) = \sigma^2/N = \sigma_s^2$$

where N is the number of individuals in the sample (Sage and Melsa, 1971, p. 212). The variance equation therefore can be written

$$\hat{\sigma}_t^2 = \hat{\sigma}_m^2 + \hat{\sigma}^2/N$$

where the "hats" indicate estimates. The estimate of population variance is based on calculated jackknife values:

$$\hat{\sigma}^2 = NJ(\hat{\sigma}_t^2) - N\hat{\sigma}_m^2$$

We use values of machine precision $(\hat{\sigma}_m)$ for $\delta^{18}O$ and $\delta^{13}O$ of 0.09% and 0.07%, respectively (Thierstein and Woodward, 1981).

Confidence limits were calculated earlier using the jackknife estimates $J(\hat{\sigma})$ and var $J(\hat{\sigma})$. Note that the population variance was jackknifed rather than the standard deviation in the case of the stable isotope data. That was done for the following reason. In order to calculate confidence limits for any parameter θ, we need the quantities $J(\hat{\theta})$ and var $J(\hat{\theta})$. When we jackknife variances, we obtain an estimate of $\hat{\sigma}_t^2$ (from which $\hat{\sigma}^2$ was obtained above) and var $\hat{\sigma}_t^2$. Since the two error sources are independent and additive, and the variance of $\hat{\sigma}_m$ (a constant) is zero,

$$\text{var } J(\hat{\sigma}_t^2) = \text{var } [\hat{\sigma}_m^2 + \hat{\sigma}/N] = 1/N^2 \text{ var } \hat{\sigma}^2$$

and

$$\text{var } \hat{\sigma}^2 = N^2 \text{ var } J(\hat{\sigma}_t^2).$$

If standard deviations are jackknifed no simple way is known to extract var $\hat{\sigma}$. (Jackknifing variances is an accepted method for computing confidence limits, e.g., Miller, 1968. We could have jackknifed variances in earlier examples and achieved similar results. Instead we followed examples in the literature.)

Results of jackknifing stable isotope data are shown in Table 2.2. Analyses were performed on samples containing 30 individual foraminifers in the size interval 350–420 μm. We can compare our intersample variability estimates with estimates of intrasample variability by using stable isotope data given in Schiffelbein and Hills (1984). Those workers measured stable isotopes on single shells and small groups of shells from a single sample (a sample was a 1 cm thick semicircular piece from a 9 cm diameter subcore) from box core ERDC 83BX, a companion core to the piston core used in this paper. Data in those authors' paper (stable isotopes for foraminifers *G. sacculifer*, 350–420 μm, and *Pulleniatina obliquiloculata*, 350–420 μm) were jackknifed using variances, and are given in Table 2.3.

For example, we examine the precision of measurement using *G. sacculifer*, 350–420 μm, 30 individuals, at the 90% confidence level. Using intrasample variability values in Table 2.3, we obtain

TABLE 2.2 Jackknife-Based Variance Estimates for Stable Isotope Data

Variable	Degrees of freedom	$J(\hat{\sigma}_J^2)$	var $J(\hat{\sigma}_J^2)$	$\hat{\sigma}^2$	$\sqrt{\text{var}\,\hat{\sigma}^2}$	Confidence limits (variance)				
						80%	90%	95%	97.5%	99%
σ^{18}O (*G. sacculifer*)	20	0.035	0.0011	0.81	0.98	1.65	2.11	2.50	2.85	3.29
δ^{18}O (*N. dutertrei*)	19	0.024	0.00034	0.48	0.55	0.95	1.21	1.43	1.63	1.88
δ^{13}C (*G. sacculifer*)	19	0.023	0.0004	0.43	0.60	0.95	1.23	1.47	1.69	1.95
σ^{13}C (*N. dutertrei*)	19	0.028	0.00076	0.58	0.83	1.29	1.68	2.02	2.32	2.69

$$\delta^{18}O: \hat{\sigma}_t^2 = \hat{\sigma}_m^2 + \hat{\sigma}^2/N = (0.09)^2 + 0.52/30 = 0.0254$$

$$\hat{\sigma}_t = 0.16$$

$$\sigma^{13}C: \sigma_t^2 = (0.07)^2 + 0.57/30 = 0.0239$$

$$\hat{\sigma}_t = 0.15$$

With Table 2.2 values (intersample variability),

$$\sigma^{18}O: \hat{\sigma}_t^2 = 0.0784$$

$$\hat{\sigma}_t = 0.28$$

$$\sigma^{13}C: \hat{\sigma}_t^2 = 0.0459$$

$$\hat{\sigma}_t = 0.21$$

We can compare in a rough way intrasample variability values for *P. obliquiloculata* with our intersample variability for *N. dutertrei,* because both species inhabit the deeper part of the photic zone and are subject to similar environmental influences. Again, using 30 individuals at the 90% confidence level, for *P. obliquiloculata*

$$\delta^{18}O: \hat{\sigma}_t^2 = 0.0131$$

$$\hat{\sigma}_t = 0.11$$

$$\delta^{13}C: \hat{\sigma}_t^2 = 0.0102$$

$$\hat{\sigma}_t = 0.10$$

For *N. dutertrei*

$$\delta^{18}O: \hat{\sigma}_t^2 = 0.0484$$

$$\hat{\sigma}_t = 0.22$$

$$\delta^{13}C: \hat{\sigma}_t^2 = 0.0609$$

$$\hat{\sigma}_t = 0.25$$

From these comparisons, stable isotope intrasample variability estimates clearly are considerably smaller than estimates of intersample variability. In the case where a direct comparison could be made *(G.sacculifer),* intersample variability is roughly 1.5 $\times$ larger than intrasample variability for the example given (a typical analysis). (*F*-test significance is 99.5% for oxygen and 95% for carbon.) Although comparison using *P. obliquiloculata* and *N. dutertrei* cannot be made rigorously, a difference in variability of

TABLE 2.3 Intrasample Stable Isotope Variability

Variable	Degrees of freedom	$\hat{\sigma}^2$	$\sqrt{(var\ \hat{\sigma}^2)}$	Confidence limits (variance)					
				80%	90%	95%	97.5%	99%
$\delta^{18}O$ (G. sacculifer)	29	0.13	0.30	0.39	0.52	0.64	0.74	0.87
$\delta^{18}O$ (Pulleniatina)	29	0.041	0.083	0.11	0.15	0.18	0.21	0.25
$\delta^{13}C$ (G. sacculifer)	29	0.18	0.30	0.44	0.57	0.69	0.79	0.92
$\delta^{13}C$ (Pulleniatina)	29	0.062	0.074	0.13	0.16	0.19	0.21	0.24

greater than a factor of two (for the example) is convincing evidence of the importance of sample size and sampling strategy in determining population variability.

CONCLUSIONS

Determination of error measures is a necessary step in proper interpretation of geologic data. An appropriate sampling frame is extremely important. In particular, the population of interest must be ascertained. We have used multiple lateral samples from a single stratigraphic horizon to estimate population variability. That approach seems preferable to determining intrasample variability since intersample variability estimates more accurately reflect data reproducibility in nature.

REFERENCES

Bendat, J. S., and Piersol, A. G., 1971, *Random data: analysis and measurement procedures:* Wiley Interscience, New York, 407 p.

Box, G. E. P., 1953, Non-normality and tests on variances: *Biometrika,* v. 40, p. 318–335.

Efron, B., 1982, *The jackknife, the bootstrap and other resampling plans:* SIAM, Philadelphia, 92 p.

Gray, H. L., and Schucany, W. R., 1972, *The generalized jackknife statistic:* Marcel Dekker, New York, 308 p.

Killingley, J. S., Johnson, R. F., and Berger, W. H., 1981, Oxygen and carbon isotopes of individual shells of planktonic foraminifera from Ontong-Java plateau, equatorial Pacific: *Palaeogeogr. Palaeoclim. Palaeoecol.,* v. 33, p. 193–204.

Miller, R. G., 1968, Jackknifing variances: *Ann Math. Stat.,* v. 35, p. 1594–1604.

Miller, R. G., 1974, The jackknife—A review: *Biometrika,* v. 61, p. 1–17.

Mosteller, F., and Tukey, J. W., 1977, *Data analysis and regression:* Addison-Wesley, Reading, Mass., 588 p.

Quenouille, M. H., 1956, Notes on bias in estimation: *Biometrika,* v. 43, p. 353–360.

Sage, A. P., and Melsa, J. L., 1971, *Estimation theory with applications to communications and control:* Robert E. Krieger, New York, 592 p.

Schiffelbein, P., 1984, Extracting the benthic mixing impulse response function: a constrained deconvolution technique: *Mar. Geol.,* v. 64, p. 313–336.

Schiffelbein, P., and Hills, S., 1984, Direct assessment of stable isotope variability in planktonic foraminifera populations: *Palaeogeogr. Palaeoclim. Palaeoecol.,* v. 48, p. 197–213.

Thierstein, H. R., and Woodward, D. L., 1981, Paleontological, stratigraphic, sedimentological, geochemical, and paleogeographical data of Upper Cretaceous and lowermost Tertiary sediment samples: *Scripps Institution of Oceanography Ref. Ser.*, v. 81–82, 85 p.

Tukey, J. W., 1958, Bias and confidence in not-quite large samples: *Ann Math Stat.*, v. 29, p. 614.

Wright, T., and Tsao, H. J., 1983, A frame on frames: an anotated bibliography, *in* Wright, T. (ed.), *Statistical methods and the improvement of data quality*, Academic Press, Orlando, Florida, p. 25–72.

Yamane, T., 1967, *Elementary sampling theory:* Prentice-Hall, Englewood Cliffs, N.J. 405 p.

3

Sorting Out Geology— Unmixing Mixtures

Robert Ehrlich
and
William E. Full

Most geologic samples actually are mixtures of simpler components. For instance, many igneous rocks are mixtures of magma types or magma and "wall rock contamination." Similarly, detrital sands are generally mixtures that represent complex contributions from many source rocks or source terranes. Indeed, many sedimentologists think that size-frequency distributions are mixtures of complex "subpopulations" related to various transport modes. Furthermore, paleontologists are aware that fossil assemblages often represent complex mixtures of various ecological assemblages swept together by waves and currents.

Our intent is to show how one can define "subpopulations" that, when mixed together in various proportions, can reproduce the original data set. Generally, if samples are considered to be mixtures, then three mixing parameters are of interest: (1) the number of components in the mixture, (2) the identity of each component, and (3) the relative proportions of each component in each sample. If the composition of each potential component is known prior to analysis, then the other parameters can often be determined on a sample-by-sample basis. Clay mineral suites are often determined by comparison to a standard set of diffraction data. But what if each component in a sample is itself an assemblage (e.g., clay mineral suites of differing provenance—that is, having the same mineral assemblage but differing in proportions—which mix at the site of sedimentation); or what if component compositions are not known (e.g., igneous whole–rock compositions)? In such cases, individual samples cannot be "unmixed" (the three mixing parameters cannot be determined). This is generally the case with respect to geological samples. However, given a *set* of samples in which the mixing proportions vary from sample to sample, it is often possible to estimate the mixing parameters from a simultaneous analysis of all samples rather than by analysis of each sample individually. In fact, the procedure relies in its first stages on algebra designed to determine the solution of sets of linear equations.

Obviously, to determine this solution the data has to contain enough information to define "subpopulations." Thus the ability to unmix mixtures is limited by what measurements are taken. For example, if a geochemical population consists of mixtures of five subpopulations, then if we measure only a single characteristic in each sample, we will not be able to define the five "subpopulations." However, if a minimum of five measurements are taken, the five "subpopulations" could be defined. The number and type of measurements that comprise a data set are therefore critical to unmixing procedures.

MULTIPLE DIMENSIONS—MULTIPLE VARIABLES

How many variables are associated with the analysis of a single sample? In some cases only one; in other cases, many. For instance, a sample from a gold mine might yield a single number—gold content. Discussion of analysis and interpretation of data from collections of such samples is the concern of conventional numerical-statistical analysis (i.e., univariate analysis).

On the other hand, a sample may be represented by many variables, such as relative proportion of major elements or trace elements in petrology, relative volumetric proportions of size class intervals of sediment, and modal proportions from petrographic point counting. In the case of some variables, such as size, each sample is quantitatively different (e.g., class interval proportions), whereas in other samples each variable may represent the amount of a qualitatively different component (e.g., chemical analysis).

These two sorts of data are more similar than generally appreciated—especially if in either case each sample is suspected of representing a mixture in the sense discussed here. A whole-rock analysis can be represented by a bar chart with each bar representing a chemical element and the height of each bar representing relative proportion of that element. Samples of the other sort also can be represented by bar charts. In this case, each "bar" represents a "class interval," with each class interval representing a portion of the range of values of a variable such as grain size. The height of each bar would represent, in the case of grain size data, the representative proportion of the sample in each class interval. One particular sort of bar chart is termed a "histogram," but histograms are only one of many ways to tabulate such data and often not the most expedient good way. In both cases, each sample is represented by several variables and data from all samples can be displayed in a table or matrix. With many variables per sample evaluated, the data set can be analyzed with reference to "unmixing."

The values of each of the variables measured from a sample can be expressed simultaneously by plotting a single point in a coordinate system wherein each reference axis represents the abundance of one of the variables. The multivariate composition of a sample is just the coordinates of the sample. The number of variables thus defines a "measurement space" with all reference axes radiating outwards from a common origin mutually perpendicular to one another. We can only visually inspect such plotted points when the number of variables is three or less. However, mathematical analysis of the relationships between sample vectors (for that is what such points represent) is just as easy whether the the vectors are embedded in a two-dimensional or 12-dimensional (12 reference axes) space. Much of the ensuing discussion will concern whether or not the sample vectors "take advantage" of the dimensionality of the measurement space. That is, whether the sample vectors spread out uniformly forming a hyperspherical cloud or, instead, are restricted to, for instance, a plane canted at some angle in the measurement space.

UNMIXING

One of the principal goals of geologists who collect samples that are mixtures should be to determine the three parameters associated with any mixture: (1) the number of components comprising the mixture (the number of things mixed together), (2) the composition of each component, and (3) the relative proportions of each of the components in each sample (the recipe for recreating each sample). These goals are well understood in igneous petrology and, unfortunately, only rarely sensed in our own specialty, detrital sedimentology. Igneous petrologists are forced both by field observation (e.g., xenoliths fading into an intrusive) as well as from the data supplied by experimental petrology (phase equilibria at various pressures) to consider that the composition of igneous rock samples are mixtures. Until recently their estimates of the mixture parameters were derived by trial and error, or by establishing components (termed "end members") on the basis of theory and then determining whether or not mixtures in various proportions could reproduce the suite of rock analyses. As we shall see below, an igneous petrologist, A. T. Miesch (1976), carried this analysis to its epitome and then made a critical logical leap, which made possible the first generalized "unmixing" algorithm.

Sedimentologists have not been as fortunate. The history of size analysis is as good an example as any. Early in the history of the field, workers realized that not only were all sand grains in a sample not of the same size, but that the ranges of sizes present varied from sample to sample. Eventually this insight ripened into the concept of the size-frequency distribution. In the case of sedimentology, this can be defined to consist of the relative proportions of the sample volume in a series of adjacent size ranges. These size ranges are known as class intervals.

BAD MOMENTS IN SEDIMENTOLOGY

How did sedimentologists usually handle grain size data? By calculation of a series of indices termed "moments." The most widely known moments in geology are the mean, standard deviation, skewness, and kurtosis. In actuality, these constitute the first four of an unlimited number of moments that, mercifully, have so far gone unnamed. In statistics moments are useful only in those cases where the entire frequency distribution (i.e., all class interval proportions) can be regenerated from knowledge of the values of the moments alone. The extent that the distribution cannot be recovered by calculations from the moments is a measure of the information lost by characterizing the distribution by moments.

Only frequency distributions that are Gaussian ("normal") or closely approach normality (unimodal, symmetrical) are fair game for efficient characterization by calculation of a small number of moments. Grain size data are neither normally nor lognormally distributed. If they were distributed in either way, then cumulative frequency distributions would plot as straight lines on the appropriate probability graph paper.

Therefore, many workers have proposed that the distributions are composite, that is, composed of several "subpopulations"—each tied to a different transport mode. Others have made the same proposal based on the less exciting observation that sediment is commonly laminated, with adjacent laminae often having differing grain sizes. Because most samples are composed of many laminae, the size-frequency distribution must be composite, with its character controlled by the relative proportions of each kind of lamina and the degree of contrast between them. Whatever the root causes of such mixtures, the consensus that size-frequency distributions are mixtures should lead (but so far has not) to a general understanding that the decomposition of such mixtures cannot be accomplished by calculating means and standard deviations.

RESOLUTION OF THE UNMIXING PROBLEM

The first step is to first determine whether or not each sample in a collection consists of a mixture of a relatively small number of components. If a sample set meets this criterion, we may then estimate the number of components, the composition of each component, and the relative proportions of each component within each sample.

In the next section we discuss the manner in which these three parameters can be estimated, given a set of samples each represented by a collection of properties. This "unmixing" approach requires two separate steps. In the first, the number of components is determined. In the second step, the end member compositions and mixing proportions are estimated. The first procedure relies on some simple application of linear algebra; the second is based on Euclidean geometry. In the following paragraphs we attempt to use a minimum of mathematical terminology or equations. However, avoidance of the Scylla of mathematics has brought us close to the Charybdis of prolixity.

DETERMINATION OF NUMBER OF COMPONENTS

The number of components in a constant-sum mixture (the sum of all sample measurements in a sample is equal to the sums in each of the other samples; that is, each and every sample sums to unity) must be equal to or less than the number of measurements made per sample. That is, in the case of whole-rock analysis, the number of components must be less than or equal to the number of chemical elements present. In the case of frequency distributions, the number of components must not exceed the number of class intervals. In addition, the number of components must not exceed the number of samples in the analysis. The number of *independent* components must not exceed the number of variables or the number of samples. In most situations, the actual number of components, or "end members," is appreciably less than these maximum values.

The fact that the correct number of end members (generally less than this maximum) can be deduced is a consequence of the fact that in all cases discussed above we are dealing with constant-sum variables. It is commonly realized that constant-sum variables are inherently correlated—not independent (Chayes, 1960). This comes about because, if the sum of k variables (such as chemical elements) total, say 100%, then as we learn values of successively more variables, the permissible values of the remaining variables are limited increasingly by the restriction that the sum of all the variables must total 100%.

This lack of statistical independence of variables can, however, be expressed in many ways. The manner in which variables are intercorrelated depends on, or contrariwise *defines,* how the collection of samples (each sample containing the same k variables) are related one to another. For instance, aluminum and silicon usually vary inversely (are negatively correlated) in igneous rock samples because aluminum can replace silicon in tetrahedral sites in the crystal lattice. If the composition of each sample is the result of the mixing of several components, then further restrictions are imposed on "allowable values" for each variable.

A large degree of dependence between variables implies that the same amount of variability can be carried by a smaller number of new variables, each of which *would* be mutually independent. The new variables (representing new coordinate reference axes) can be defined and related to the original variables via linear equations. Such operations have sometimes been termed *vector analysis* or *factor analysis*—the last appellation carrying rather more heat than light.

The unmixing operation can be called vector analysis because each sample can be considered a *multidimensional* vector, with each original variable scaled on separate reference axes at right angles to one another, called *orthogonal axes.* The direction cosines (set of cosines of the angle between any two vectors and the origin of the measurement system) between all possible pairs of sample vectors constitute a sort of correlation matrix sometimes called a *similarity* matrix. From this square matrix, or its mathematical equivalent, the number of new independent variables can be determined and defined. In such an analysis, new variables are represented by new reference axes. Each new variable accounts for a certain proportion of the total variance in the system and is represented by an *eigenvalue.*

Resolution of the unmixing problem lies in the fact that when samples measured according to a constant-sum measurement system are, in fact, mixtures of components, then *the number of eigenvalues equals the number of components.* Thus, if three eigenvalues account for all the variance, the underlying geologic reality can only result from, at most, a three-component system. That is, even though we may have started with many variables (for example, 50), the sample vectors are actually confined to a three-dimensional rosette. The three reference axes are themselves vectors and are termed *eigenvectors.* Each eigenvalue is associated with a corresponding eigenvector.

In fact, however, the three (in this case) eigenvalues may not account for strictly *all* of the variance. This happens because in any measurement system there exists a component of variance arising from a natural background "noise" including analytical error. The problem then is to set criteria for deciding the amount of the total variance of the system that would be left unexplained.

Initially, as many eigenvalues are calculated as there are original variables. However, if ranked in descending order of variance accounted for, often the first few eigenvectors account for most of the variance. A major problem, however, is to determine exactly when to stop accumulating eigenvalues (i.e., how much variance is essentially "all"?).

AN ASIDE ABOUT THE RELATIONSHIP BETWEEN VARIANCE AND INFORMATION

In the framework of our discussion, we mentioned that each eigenvalue represents a portion of the total variance of the system. If eigenvalues are arranged in order of the decreasing variance accounted for (the usual case), can we say that the first eigenvalue is most important because it is associated with the most variance? The answer is, of course, no.

Consider once again our exemplary three-component system. If samples consist of mixtures of basalt, granite, and partially melted metashale, then changes in the amount of the basaltic component, with respect to the granitic, will result in a greater intersample variance than if the proportion of granite to metashale underwent a change of similar magnitude. The reason for this is simply that compositions of granite and metashale are compositionally closer to one another than either is to basalt.

However, the basalt component cannot be considered *inherently* more important than the other components in the mixture. A common erroneous statement made by ignorant practitioners of factor analysis is that "only the first k eigenvalues (k less than the number of original eigenvalues) will be considered inasmuch as they account for 70% of the variance."

SUCCESSFUL UNMIXING

Successful solution of the constant-sum unmixing problem is due largely to the systematic investigations of Miesch (1976). Full et al. (1981, 1982) generalized Miesch's work to extend its usefulness. Miesch's formulation arose directly from his interest in understanding whole-rock, major-element compositions of igneous rocks. He recognized that two sorts of solutions were appropriate in his personal research; both relate to the theoretical framework of igneous petrology and both relate directly to hypothesis testing.

In the first case, the petrologist collects a suite of samples including those representing *potential* "pure end members," such as chilled margins of plutons or carefully selected dike rocks. In the second case, *potential* end members could be predicted on the basis of phase-equilibria studies; these might be termed "theoretical end members." Miesch, building on the work of Imbrie (1963) and Klovan and Imbrie (1971), saw that two different analytical steps were required in order to solve the problem; the number of end members must be defined and, once determined, another sort of analysis would provide tests of the feasibility of his end member models.

The first step simply involved determination of the number of eigenvalues from the directional cosine similarity matrix, or its mathematical equivalent. Miesch (1976) devised four criteria for estimating the correct number of eigenvalues—two explicit and two implicit. The first criterion is that most of the variance in the data ensemble should be preserved. In such a case, relations between samples (sample vectors) are preserved. In most cases, more than 90% of the variance is represented by a small number of eigenvalues. However, not *all* variance is represented—some must be "set aside" for random variation. What are the consequences of selecting k eigenvalues (accounting, say, for 90% of the variance) rather than $k + 1$ eigenvalues (accounting for 93% of the variance)? Here Miesch uses a second criterion.

If all variance is not accounted for, the residual variance might be spread "evenly" across all of the originally measured variables or might be concentrated on only a few. To determine the true situation, Miesch proposed back-calculating from eigenvector space (fewer axes) to the original variable space (more reference axes). If values of a measured variable (MgO, for instance) can be recovered with good precision, then one can state that this variable supports a k-component solution. If, however, a given variable cannot be back-calculated with precision (i.e., has a small coefficient of determination when compared to its actual value), then one can say that the k-component solution is not supported by that variable. The analyst then judges, *based on experience and theory,* whether the variable(s) that do *not* support the analysis are important (Figure 3.1).

These criteria are embedded in the algorithm "EXTENDED CABFAC" of Klovan and Miesch (1976). Implicit in the computer program "EXTENDED CABFAC" (EC) is another criterion not explicitly discussed by Miesch (1976) or Kovan and Miesch (1976). This is related to another common source of variance—the *outlier* or *maverick* sample. Such samples may represent erroneous encoding of data, erroneous analytical procedure, or a sample from a population separate from the one of interest. Of course, it might represent a relevant, albeit extreme, sample. Such a judgment can only be made by the investigator within the context of an investigation.

Inspection of normalized varimax factor loadings in EXTENDED CABFAC can help detect outliers. Varimax factor loadings display the relative relationship between each sample and the eigenvectors related to the eigenvalues. Often, many of the higher eigenvectors, accounting for little variance, contain loadings that are small for all samples (e.g., <0.001). The odds are that such eigenvectors express variance associated with random "noise." Alternatively, in some cases, most loadings will be essen-

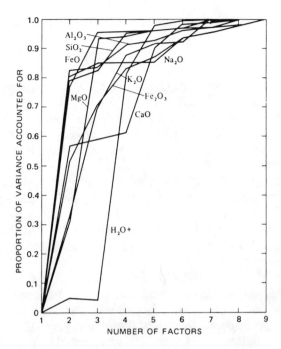

Figure 3.1 Use of the coefficient of determination (proportion of variance accounted for by each variable) as a way to determine the number of end members. A two-component system is inconsistent with respect to calcium, sodium, iron, and water (Skaergaard layer series from Miesch, 1976).

tially zero, but one or two *samples* will have relatively large loadings. These are outliers.

If outliers are judged not to be critical to the analysis, then fewer eigenvectors are necessary. However, the analyst *must* remember that using fewer eigenvectors (defining the system in terms of fewer end members) will distort the relationship of such outliers with respect to other samples.

The fourth criterion for determination of the number of components is "scientific reasonableness," which will be discussed in the context of the algorithm QMODEL.

DETERMINATION OF END MEMBER COMPOSITIONS AND MIXING PROPORTIONS—THE QMODEL FAMILY OF ALGORITHMS

Using criteria discussed above, a candidate set of eigenvectors can be determined. In the case of constant-sum data, the number of eigenvectors calculated in EXTENDED CABFAC equals the number of components in the mixing system. The QMODEL family of programs relate samples (i.e., mixtures) to end members in various ways.

QMODEL (Klovan and Miesch, 1976) in its original formulation was used to test petrologic models. QMODEL (QM) and its extensions by Full et al. (1981, 1982)—EXTENDED QMODEL (EQM) and FUZZY QMODEL (FQM)—are not in any sense part of factor analysis. First, in factor analysis, the *scientific* importance of an eigenvector is generally believed to be directly tied to the amount of variance it represents. Second, each eigenvector (termed a "factor") is supposed to, in and of itself, represent some underlying real property. That is, factor analysis is supposed to "explain" a variance–covariance matrix. However, explicability depends to a large part on an assumption of a homogeneous normally distributed variance. In the QMODEL procedure, described below, the eigenvectors play a passive role as a set of coordinate axes and no special importance is placed on one or another.

The object of the QMODEL programs is to enclose data vectors within a straight sided geometric figure (called a *polytope*) whose number of sides (or edges) is equal to the dimensionality of the basis space (number of eigenvectors) plus one. In this case all samples (vectors) can be described as linear combinations of compositions represented by each vertex on such a figure. If this is done "correctly," then the other two parameters of constant-sum unmixing—end member composition and mixing proportions—are easily accomplished.

Because essentially all of the variance had been preserved in going from a system of greater dimensions (one axis for each measured variable) to a space of lesser dimensions, the *relations between sample vectors are the same in both spaces.*

If the system has three components, the enclosing reference figure will be a triangle (not necessarily equilateral). A four component system demands a tetrahedron. Generalizing an n component system must involve an n-1 dimensional solid.

QMODEL

In the original QMODEL (Miesch, 1976), vertices of the polytope were defined in one of two ways. In one way, the n (where "n" is the number of included eigenvectors) samples mutually farthest apart are defined as vertices and mixing proportions of other samples calculated (Figure 3.2). If one or more samples is represented by one or more negative proportions (e.g., -25% Mg), then samples at polytope vertices cannot represent true end members. If end member proportions of all samples are nonnegative, the n most extreme samples are, at the least, *feasible* end members.

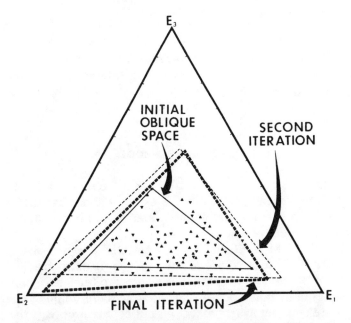

Figure 3.2 Results of the iterative solution in EQM providing positive mixing proportions. The initial QM solution based on extremes in the data (initial "oblique space") was used as a starting position for the EQM procedure. Vertices of the triangular polytope defined by the final iteration represent end members with nonnegative compositions (Figure 6 from Full et al., 1981).

In the second case, vertices are defined by a set of *n* compositions defined by theory. These do not represent real samples but compositions that are realistic in that all proportions of components *within* each end member are either zero or positive. Again, if negative mixing proportions occur, some samples vectors fall outside the polytope volume and some or all of the theoretical end members, represented as polytope vertices, are not feasible.

Feasibility or unfeasibility of the QMODEL is thus a way of testing the sufficiency of petrogenetic models based on field criteria or on thermodynamic considerations. A feasible result does not, however, ensure that end members necessarily represent end members of the natural system. For example, one might have included an additional eigenvector in the analysis, for instance one more end member. Here we have another implicit criterion of the QMODEL algorithm for determining the number of end members. What usually is done is to calculate a *series* of solutions that involve one or two dimensions smaller and larger than *n*. After inspecting loadings for outliers and rejecting solutions where many large loadings occur on eigenvectors greater than the greatest included in the analysis, results in terms of end member compositions and mixing proportions are evaluated within a scientific context. That is, of course, if samples were collected (hopefully!) in terms of some spatial or chronological framework. Do the mixing proportions make sense in this framework? Are end member compositions acceptable in terms of theory? In this light one solution may be more suitable than another.

EXTENDING QMODEL

The QMODEL algorithm is successful within the framework for which it is intended. However, is it possible to extend the algorithm to calculate feasible end members that are neither captured within the sample set nor predicted theoretically?

EXTENDED QMODEL and FUZZY QMODEL were designed by Full et al. (1981, 1982) to accomplish this. In one of its modes, QMODEL assumes that the n mutually extreme samples are true end members, and that if the resulting polytope truly encloses the set of sample vectors, then at least those end member samples will not be mixtures (e.g., the proportions of all other end members will be zero for each end member sample).

If these extreme samples are not feasible end members, then EQM-FQM takes over. Both of these algorithms assume that at least one sample will occupy either a vertex or lie along one face edge of the polytope. In this case, at least n samples contain zero abundance of *at least* one end member. Both EQM and FQM are iterative procedures that, starting with an initial polytope, progressively enlarge and rotate it until convergence occurs. They differ from one another in how the initial polytope is determined.

EQM-FQM

EXTENDED QMODEL is the simplest extension of the original QMODEL algorithm/program of Klovan and Miesch. It commences by determining, or trying to determine, the n most mutually extreme samples. Next it determines the polytope and, then, if all samples are enclosed, terminates (this *is* the original QMODEL). However, if negative mixing proportions exist (samples lying outside the polytope), it moves the polytope edge outwards, parallel to itself, a stated distance or until it "touches" the outermost sample. At this point all samples have nonnegative mixing proportions and one or more vertices of the polytope (representing end member compositions) are no longer located on a sample. Such vertices represent potential samples more extreme than those occurring in the sample array.

However, often it is the case that part or all of the composition of one or more end members will be negative. This indicates that one or more vertex has strayed into the negative orthant (an *orthant* is the multidimensional equivalent of a quadrant in two dimensions) in our variable space. When this occurs, the program "zeros" out the negative component. This essentially forces the polytope to rotate slightly (i.e., the edges no longer parallel to QMODEL initial polytope). When this occurs some samples may stray outside the new polytope, again producing negative mixing proportions. If this happens, reiteration occurs.

FUZZY QMODEL follows the same general procedure except that the initial polytope is defined by the general *n*-dimensional shape of the sample cloud. The efficacy of one or the other is discussed in Full et al. (1982).

QUALITY OF UNMIXING

We have used the QMODEL algorithms several hundred times in the past three years. They have produced results of enough quality to provide genuinely deeper insights into the nature and genesis of geological systems. They have been used to "unmix" organic gas chromatographic data (Kornder and Carpenter, 1984), sand grain provenance (Mazzullo and Ehrlich, 1983; Kennedy and Ehrlich, 1985) and whole-rock chemistry (Horkowitz, Stakes, Shervais and Ehrlich, in prep.). Currently we are using it to analyze microfossil assemblages and to classify pores in reservoir rocks (Ehrlich et al., 1984).

The technique is, however, only as good as the quality of the sample set because it relies on innate relationships between samples located as vectors in *n* space. The more biased the sampling plan, the worse the results are. In other words, try to collect a set of samples that expresses the spectrum of variability of the system of interest. In our experience, end members must vary by at least 5% in the mixture in order for end members and mixing proportions to be estimated.

Finally, a nonconstant-sum version (useful for trace elements and water analysis) has been developed and is currently undergoing testing.

REFERENCES

Chayes, F., 1960, On correlation between variables of constant sum, *Jour. Geophys. Res.*, v. 65, no. 12, p. 4185–4193.
Ehrlich, R., Kennedy, S. K., Crabtree, S. J. and Cannon, R. L., 1984, Petrographic image analysis I—Analysis of reservoir pore complexes, *Jour. Sed. Pet.*, v. 54, no. 4, p. 1365–1376.

Full, W. E., Ehrlich, R., and Klovan, J. E., 1981, EXTENDED QMODEL—Objective definition of external end members in the analysis of mixtures, *Jour. Math. Geol.*, v. 13, no. 4, p. 331–344.

Full, W. E., Ehrlich, R., and Bezdek, J. C., 1982, FUZZY QMODEL: A new approach for linear unmixing, *Jour. Math. Geol.*, v. 14, no. 3, p. 259–270.

Horkowitz, J. P., Stakes, D. S., Sheravis, J. W., and Ehrlich, R., 1987, Fundamental differences in magma chamber processes between basalts from the FAMOUS and AMAR rifts, mid-Atlantic ridge, in prep.

Imbrie, J., 1963, *Factor and vector analysis programs for analyzing geologic data,* Office of Naval Research, Geography Branch, Tech. Rept. 6 (ONR Task No. 389–135), 83 p.

Kennedy, S. K., and Ehrlich, R., 1985, Origin of shape changes of sand and silt in a high-gradient stream system, *Jour. Sed. Pet.*, v. 55, no. 1, p. 57–64.

Klovan, J. E., and Imbrie, J., 1971, An algorithm and FORTRAN IV program for large scale Q-mode factor analysis and calculation of factor scores, *Jour. Math. Geol.*, v. 3, no. 1, p. 61–76.

Klovan, J. E., and Miesch, A. T., 1976, EXTENDED CABFAC and QMODEL Computer programs for Q-mode factor analysis of compositional data, *Comput. Geosci.*, v. 1, p. 161–178.

Kornder, S. C., and Carpenter, J. R., 1984, Application of a linear unmixing algorithm to the normal alkane patterns from Recent salt marsh sediments, *Org. Geochem.*, v. 7, no. 1, p. 61–71.

Mazzullo, J., and Ehrlich, R., 1983, Grain shape variation in the St. Peter Sandstone: A record of eolian and fluvial sedimentation of an early Paleozoic cratonic sheet sand, *Jour. Sed. Pet.*, v. 53, no. 1, p. 105–119.

Miesch, A. T., 1976, Q-mode factor analysis of geochemical and petrologic data matrices with constant row-sums. *Geol. Surv. Prof. Paper 574-G*, 47 p.

4

Consistency of the Two-Group Discriminant Function in Repartitioning Rocks by Name

Felix Chayes

The ease with which multivariate reductions can now be performed is one of the major advantages and hazards of contemporary statistical geology. This chapter is concerned with one particular species of the genre, the two-group discriminant function in its simplest form. The basic theory of discriminant function analysis, like that of most forms of multivariate reduction, was completed before the start of World War II. For many years there were no applications of the procedure in practical statistical work in the geosciences and few in the biological sciences, probably because to naturalists, for whom it should have had maximum appeal, the amount of calculation required seemed disproportionate to the expected yield. Even early electronic computers could be programmed to perform calculations of this sort expeditiously, however, and the great boom in applied multivariate data reduction in the naturalistic sciences starts with the resumption of normal scientific work at the end of World War II.

In recent years there have been numerous attempts to standardize the procedures by which petrologists classify volcanic rocks. In large part this activity consists of replacing qualitative criteria based on petrography and mineralogy by quantitative ones based on bulk chemical composition. The IUGS group responsible for the work is by no means committed to formal statistical treatment of the problem—quite the opposite, in fact—but my own inclinations make it impossible for me to resist attempting just that, as a private exercise. The Fisher discriminant function (Fisher, 1936; see also Morrison, 1967, or Kendall and Stuart, 1976) is an obvious candidate for consideration in this respect. A major strength and a major weakness of this form of reduction on compositional data are discussed below, using an ad hoc petrographic example. In forming a two-group discriminant one starts from a prior classification of a set of objects and generates a linear function, based on variables other than those used in the original classification, in such fashion that the "scores" calculated from it maximize the ratio of between- to within-group sums of squares (of deviations).

In replacing petrographic and mineralogical criteria by a bulk-chemical sample statistic, it is not our intention to undo or abandon the taxonomic work satisfactorily performed by those criteria; rather, our objective is to extend the existing classification to objects that cannot be classified by means of them. That this is in fact our primary concern is clearly illustrated by the practice of using the same class names in the prior (petrographical–mineralogical) and posterior (bulk chemical) classifications. A high level of consistency* of the prior and posterior classifications is thus a major requirement—the alternative is simply continuance, or perhaps even exacerbation, of the current scholarly chaos we are attempting to eliminate.

For the two-class case the distribution of class frequencies may be shown in a 2×2 contingency table,

		Prior frequencies		
		A	B	Σ_r
Posterior	A	N_1	N_2	$N_1 + N_2$
Frequencies	B	N_3	N_4	$N_3 + N_4$
	Σ_c	$N_1 + N_3$	$N_2 + N_4$	N

in which the row and column captions are class names and the N's the frequencies of their cooccurrences. What we would like to do is maximize the ratio $(N_1 + N_4)/N$. Does the discriminant function automatically maximize this ratio for us? Although in these circumstances an effective discriminant function does indeed yield a high value for what is here termed consistency, there seems no a priori requirement that it actually *maximize* the ratio $(N_1 + N_4)/N$. And, in practice, if the posterior classification is

---

*Throughout this note "consistency" has the general sense of "correspondence" or "agreement," not the statistical sense of convergence on a true value with unlimited increase in sample size. Further, the agreement in question concerns only the *proportion* of like classifications per sample, not the like classification of individual items.

bivariate, it is often possible to find, by inspection of the scatter diagram, another linear field boundary that does a little better on the data actually used in the calculation. Hence, in the spirit of this book, the application we are about to describe would perhaps be better characterized as a modest misapplication.

A few words about the rationale of the procedure, illustrated by a hypothetical two-variable two-group discriminant, may be helpful to many readers. Let us suppose someone has collected all or a large part of the published analyses of the Cenozoic volcanics of some region, and that from these we have culled the analyses of all rocks named, in the source references, either rhyolite or trachyte. We could then plot, for example, the P_2O_5 content of each analysis along one axis of a conventional scatter diagram, and its content of some other oxide, say TiO_2, along the other. If all the rhyolites fell in one tight cluster, all the trachytes in another, and the distance between the clusters were larger than the sum of their radii, the discriminant function calculated from the data would plot somewhere in the intervening space, and would in fact partition the data in complete consistency with the prior classification. But so, too, would any other line drawn through this space, and if the intercluster distance were large in relation to the sum of the cluster radii, lines yielding similarly consistent partitions might vary widely in intercept and slope. In the sense of common language, though certainly not in statistical terms, *each* of these lines is the trace of a perfectly consistent discriminant (and that, alas, is the way the term is sometimes misused in geological and geochemical studies).

As the shapes of the clusters depart from circularity and the ratio of intercluster distance to cluster radii decreases, both the region available for lines making perfectly consistent partitions and the permissible variation among the parameters of such lines decrease rapidly. Nevertheless, if a straight line could indeed partition the data with complete consistency, we could always locate it by inspection and, from this point of view, would have no need at all for computation. Even in the bivariate case, however, the only one in which appraisal by inspection is possible, groups mutually exclusive in the prior (mineralogical–petrographic) classification are usually not mutually exclusive in any posterior (bulk chemical) classification we can devise. Rather, overlap that is at least marginal and often considerably more than marginal is the order of the day. Complete consistency in the discriminant thus becomes impossible, and we are obliged to choose between discriminants all of which are to some extent inconsistent with regard to the prior classification.

To examine the possibilities with reasonable efficiency we must then rely on computation. Although at first sight rather surprising, what emerges from computation in the present case is actually not unreasonable and may prove of rather broad interest in geochemical applications of certain other multivariate reduction procedures.

As the emphasis in this book is on methods, let us use an example in which no one need feel called upon to defend any considerable investment of either credulity or credibility. Let us define the prior classification by the names applied to rocks in the source references from which the analyses were drawn. These names are not casually applied; each denotes an object for which a full bulk analysis is presented in the source reference, usually accompanied by a rather extensive mineralogical and petrographic description. They may be based partly and rather inconsistently on the amounts of particular "essential oxides" but are more likely to involve a combination of mineralogical, petrographic, and normative criteria. The posterior classification will be based on linear functions of certain of the major oxides recorded in the analyses.

Specifically, let us consider the performance of discriminants based on all combinations of SiO_2, CaO, MgO, and FeO, as devices for distinguishing from each other the basalts and andesites of a particular part of the globe. Except for SiO_2, which has been an intermittent and rather unsystematic part of the prior classification for more than a half century, this is virgin ground. Basalts and andesites are the commonest of all volcanic rocks, so distinguishing between them is a common assignment. So far as I know, however, no one has proposed that in distinguishing between andesite and basalt by means of essential oxide content one can do about as well without silica as with it. And that happens to be the case.

The data pool, drawn from file RKOC76 (Chayes et al., 1977), consists of 860 analyses of andesite and 320 of basalt, this being the complete stock currently available to me, in machine readable form, of published information concerning specimens of the two types that (a) are from Japan and (b) contain less than 2% H_2O. For each calculation or test a subsample of approximately 50 specimens was drawn randomly from each group. Each result quoted is the mean or standard deviation found for 300 such paired subsamples. In the sampling procedure used, no analysis may occur more than once in any sample, but the same analysis may occur in more than one sample. When I began this work I was concerned that the occurrence of common elements in successive samples would impose a spurious agreement on results; accordingly, the number of iterations was kept small enough (15) so that the incidence of common elements was negligible. Current statistical work (Efron, 1982; for an elementary introduction see Diaconis and Efron, 1983), however, favors the use of very large numbers of iterations under entirely comparable circumstances. Accordingly, all sampling and calculation were repeated with iterations of 300 instead of the 15 used initially; differences in observed consistencies for samples of 15 and 300 were in all cases small and in all but one case considerably less than the associated standard errors. Results shown in all tables in this chapter are for 300 iterations of samples containing approximately 50 items of each type.

TABLE 4.1 Percent Consistency of One- and Two-Variable Discriminant Functions

	SiO_2	CaO	MgO	FeO
SiO_2	89.2	90.1	90.1	91.6
CaO		82.9	89.3	84.4
MgO			80.9	87.9
FeO				77.2

To evaluate consistency, a discriminant function was generated from an initial randomly chosen pair of subsamples of approximate size 50 each, and tested on the next 300. Results for the oxides, individually and in pairs, are shown in Table 4.1. As Hatch and Wells (1926) observed more than a half century ago, SiO_2 itself is indeed a remarkably effective discriminant. Each of the other oxides of the group is much less effective individually and only one of the three pairs formed from them matches the consistency of SiO_2 alone.

Similar information is shown in Table 4.2 for discriminants based on all four of these oxides and on all combinations of three. An old-fashioned variance analysis indicates that ternary combinations do not differ significantly from each other in discriminating power but are superior to binary combinations. Petrologists will be interested to note that on this particular data set and for this particular purpose the combination CaO–MgO–FeO is superior to SiO_2 alone and comparable in discriminating power to any binary or ternary combination containing SiO_2.

Stability of the partitions, as shown by their standard deviations, is remarkable. (Although not apparent in Table 4.2, stability seems to be rather insensitive to consistency; functions with consistency barely above background—for example, Fe_2O_3–P_2O_5–Al_2O_3, with consistency of only 57%—are often only marginally less stable than the tabled entries.) In view of the stability of the partitions they yield, is it not reasonable to expect

TABLE 4.2 Percent Consistency of Three- and Four-Variable Discriminant Functions

Variables	Average	Standard deviation
SiO_2–CaO–MgO–FeO	92.7	2.4
SiO_2–CaO–MgO	91.0	2.6
SiO_2–CaO–FeO	91.6	2.7
SiO_2–MgO–FeO	92.4	2.4
CaO–MgO–FeO	91.2	2.7

TABLE 4.3 Means and Standard Deviations for
Coefficients of Discriminant Functions Generated
from 300 Subsamples of the Andesite–Basalt Pool

Coefficient	Mean	Standard deviation
A. Based on (CaO, MgO, FeO)		
CaO	0.842	0.226
MgO	0.897	0.183
FeO	0.531	0.218
Constant	15.129	2.513
Consistency	0.912	0.027
B. Based on (SiO, CaO, MgO, FeO)		
SiO_2	0.427	0.296
CaO	0.594	0.474
MgO	−0.710	0.475
FeO	−0.594	0.391
Constant	10.130	16.453
Consistency	0.926	0.024

that the functions themselves are similarly stable? In a deterministic situation most of us would indeed suppose that as they seem to behave very similarly they must also be very similar algebraically. Information bearing on this matter can be obtained by a minor modification of the procedure used to test consistency.

Instead of testing an existing function on 300 successive pairs of subsamples, we can as easily compute a new function from each of the 300 pairs. For the combination CaO-MgO-FeO the result of this procedure is summarized in Table 4.3A. Analogous information for the four-variable discriminant SiO_2-CaO-MgO-FeO is shown in Table 4.3B. In both cases the variability of consistency is markedly less than that of the coefficients, but the discrepancy appears to be much greater for the four-variable function. This is the result mentioned above as at first sight surprising but perhaps not basically unreasonable. Scaling factors are a continual nuisance in this work and, if 300 functions of a set differed from each other primarily by scaling factors, the sample variances of coefficients could well be large; the *covariances* of the coefficients would then also be large, however, and except for those involving the constant term, λ_0, the observed correlations between coefficients are mostly small. Inclusion in the discriminant function of a variable indifferent to the prior classification can only add noise to the sample statistics. Here, however, the new variable is actually *more* sensitive to the prior classification than any of the other three, and by a rather considerable margin, as indicated by the diagonal elements of Table 4.1. Evidently *any* three of these four variables provide close to the maximum taxonomic consistency possible in this data set. The fourth—whichever is the fourth and whatever its individual discriminating power—then simply contributes noise.

In summary, it is easy to construct hypothetical scatter diagrams that would be amenable to perfectly consistent partition by two-group discriminant functions with very different coefficients. If the data are still in well-defined clusters but intercluster distance is less than the sum of the cluster radii, there will be marginal overlap, and lines passing through the region of overlap may then partition the data with stable but *im*perfect consistency. Depending on cluster shape and penetration, the parameters of such lines of stable consistency could vary considerably. The curious combination of stable consistency and unstable coefficients may indeed persist even when overlap is so pervasive that satisfactorily consistent reclassification proves quite impossible. To date, this has been noted only in compositional data. It may be an indirect consequence of closure, but probably is not confined to closed data. What is at issue is the covariance structure of the array. Perhaps closure predisposes covariances to take the appropriate values, but these surely could be generated in other ways.

Symposia like ours ought to have morals, and the moral of this chapter is that multivariate functions that do admirably the work for which they are designed may not be of much use for other purposes. On petrochemical data, for instance, the two-group discriminant function seems a splendid way to construct satisfactorily stable alternative classifications, but we must not put much credence in the equations of the indicated field boundaries, and this is especially so if the calculations leading to them include variables that either do not or cannot contribute to the power of the discriminant. It is arguable, and from time to time has indeed been argued or tacitly assumed, that such many-variable functions may be approximations of fundamental-relations that are physically or petrologically interpretable. The work reported here, however, indicates they may also be extraordinarily noisy approximations, so noisy that physical or petrological interpretation of them is not likely to be of much use. Simulation work now in progress indicates that, in principal component analysis of major element compositional data, a similar relation often holds between the consistency of eigenvalues and the stability of the coefficients of related eigenvectors. Probably no substantive interpretation of an empirical multivariate function computed from compositional data should be attempted unless its coefficients have been found to be reasonably stable. Simulation, along lines described here, will usually be the most economical way in which to evaluate the stability of such coefficients.

REFERENCES

Chayes, F., McCammon, C., Trochimczyk, J., and Velde, D., 1977, Rock information system RKOC76, *in* Annual report of the director of the geophysical laboratory, Carnegie Institute of Washington Year Book 76, p. 637–638.

Diaconis, P., and Efron, B., 1983, Computer-intensive methods in statistics: *Scientific American,* v. 248, p. 116–131.

Efron, B., 1982, *The jacknife, the bootstrap and other resampling plans, Monograph 38:* Society for Industrial and Applied Mathematics, Philadelphia.

Fisher, R. A., 1936, The use of multiple measurements in taxonomic problems: *Ann. Eugenics,* v. VII, pt. II, p. 179–188.

Hatch., F., and Wells, A. K., 1926, *Petrology of the igneous rocks,* 8th ed., G. Allen and Unwin, London, p. 151–153, 248.

Kendall, M., and Stuart, A., 1976, *The Advanced theory of statistics,* v. 3, Hafner, New York, p. 327–353.

Morrison, D. F., 1967, *Multivariate statistical methods,* McGraw-Hill, New York, p. 130–133.

5

Quantitative Recognition of Granitoid Suites Within Batholiths and Other Igneous Assemblages

E. H. Timothy Whitten
Gongshi Li
Theodore J. Bornhorst
Peter Christenson
and Darrell L. Hicks

Granitoid batholiths comprise numerous discrete plutons, each of which characteristically has mappable boundaries and distinctive chemical, mineralogical, and textural features. The term *suite* has been used informally for over half a century; Bayly (1968) defined it as a "group of rocks whose field relations and compositional characteristics make it appear that they have a common source." Griffin et al. (1977), Chappell (1978; 1984), and White and Chappell (1983) claimed that chemical similarity permits individual groups of Paleozoic plutons within the Lachlan Fold Belt (S.E. Australia) to be grouped into suites, with the implication that each suite bears a geochemical signature of the distinctive parental crustal material from which it evolved by (partial) melting. Chappell (1984, p. 695) stated "that many plutons can be grouped together [as suites] on the basis of shared petrographic, chemical and isotopic features." Chappell, White, and coworkers (e.g., Beams 1980) made extensive use of the suite concept in elaborating the petrography and petrogenesis of Lachlan Fold Belt granitoids (Figure 5.1). In important reviews of world granitoid complexes, Pitcher (1982; 1983) embraced and extended these concepts. Although this suite

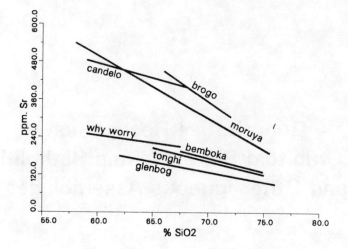

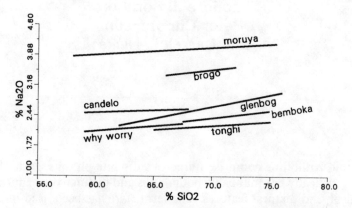

Figure 5.1 Regression lines on Harker-type diagrams based on Sr, Na₂O, and SiO₂ analyses for all samples assigned to certain named suites in Bega and Moruya Batholiths, Lachlan Fold Belt, S.E. Australia by Chappell and co-workers (after Beams, 1980).

concept does not seem to have been challenged openly in print, several petrologists have voiced scepticism about the reality and recognizability of such suites within granitoid batholiths.

The objective quantitative recognition of natural groups (e.g., suites) from numerical data for an arbitrary set of observed variables is a classic and difficult task common to many areas of the earth sciences. Whitten and Chappell (1984) and Whitten (1985) showed that cluster and discriminant analyses, based on 32 major and minor chemical components for some 304 samples, appear to corroborate the reality of many of Chappell's and Beams' named suites; these statistical methods have certain inadequacies that are now reviewed. In this chapter, an attempt is made to assess more rigorously the validity of suites (and particularly those previously identified by Chappell and Beams) within some batholiths of the Lachlan Fold Belt using Fisher's exact-test (randomization-test) approach. Although initially extremely sceptical of the objective reality of such suites, we conclude that they are probably real within the Lachlan Fold Belt, *provided* that the definition of suite is limited in a special way; some difficult problems remain and some of the named suites appear to need revision.

PREVIOUS ANALYSES

Whitten and Chappell (1984) and Whitten (1985) used whole-rock analyses for 32 elements (including FeO and Fe_2O_3) available for 304 samples representing 102 named granitoid plutons of the Bega, Gabo, and Moruya Batholiths, Lachlan Fold Belt, S.E. Australia; see acknowledgment section at the end of the chapter for the source of data. These plutons range in mapped surface area from 970.0 to 0.1 km^2. Earlier, Beams and Chappell (Beams, 1980; Beams, Chappell, and White, in preparation) had assigned almost 70 percent of these samples to some 23 named suites; in four cases, only two samples were assigned to a suite, whereas the Glenbog suite was represented by 43 samples from 14 plutons ranging in surface area from 335.0 to 7.8 km^2. Cluster analysis, based on the whole data set, resulted in clusters of samples with many similarities to previously named suites. Table 5.1 lists 18 original suites (numbered) recognized by Chappell (White and Chappell, 1983; Beams, Chappell, and White, in preparation) and four of the largest additional suites (A through D) named by Beams (1980). "Good Suites" in Table 5.1 means that Whitten and Chappell's (1984) and Whitten's (1985) cluster analyses suggested that these pre-identified suites are real (i.e., all, or almost all, of the samples previously assigned to these suites were so assigned by cluster analysis). Four "Composite Suites" comprise discrete sets of samples, but cluster analysis suggested that each should be divided into two or more sets (suites). Previously named suites listed as "Not Suites" (Table 5.1) comprise samples that do not yield cohesive sets by cluster analyses. These cluster-analysis results suggested that 10 of the original named suites are real natural groups, that four of the original suites should be subdivided, and that eight were not well defined and require reexamination. The fact that so many suites are not "Good Suites" (Table 5.1) is a cause for concern.

Conclusions reflected in Table 5.1 depend on the propriety of using cluster analysis for this purpose. Whitten and Chappell (1984) and Whitten (1985) used the BMDP Cluster Analysis computer program P2M (Dixon,

TABLE 5.1 Synopsis of Cluster Analysis of Bega, Gabo, and Moruya Batholith Granitoid Samples, Lachlan Fold Belt, Australia

Good suites			Composite suites			Not suites		
Suite number	Suite name	No. of samples	Suite number	Suite name	No. of samples	Suite number	Suite name	No. of samples
2	Xmas	2	5	Cobargo	13	1	Moruya	17
3	Gabo Island	9	8	Candelo	12	7	Warri	2
4	Mumbulla	8	10	Bemboka	39	11	Drummer	2
A	Brogo	7	13	Glenbog	43	12	Coolangubra	6
B	Wangerabell	3				14	Kybean?	2
C	Kameruka	10				15	Tonghi	15
6	Wallagaraugh	14				17	Rock Flat	3
D	Braidwood	12				18	Rossi	2
9	Why Worry	13						
16	Bukalong	3						

Source: From Whitten (1985, Table 1).

1981). The program forms clusters of samples in n-space when n variables are considered simultaneously for each sample. Cartesian east and north geographical coordinates were included with chemical components as independent variables. Each variable was normalized prior to clustering in order that each have equal weight (influence) in analyses.

Some workers (e.g., Donald Saari, personal communication, 1983) have contended that cluster analyses can be expected to yield misleading (or spurious) results for constant-sum data (i.e., data whose sum for each sample is constant, as with percentage data). Certainly, the individual chemical percentage variables used here are not independent (Chayes, 1971). Use of trace-element data alone by Whitten and Chappell (1984) and Whitten (1985) yielded almost identical results to those for all of the chemical data, or for major elements alone. Whether trace elements are constrained like major-element percentage data has been debated, but some trace elements are strongly correlated with particular major elements and thus can not be wholly independent from each other (cf. Chayes and Kruskal, 1966; Miesch, 1969). Aitchison (e.g., 1984) suggested solutions to the constant-sum (or closure) problem for various statistical analyses of geochemical data but did not address cluster analysis.

Assigning equal weight (by normalization) to each variable was a matter of convenience; thereby, each variable has equal importance or significance in the analysis. Both Chappell and Beams put particular emphasis on SiO_2, Na_2O, and Sr. Different geologists might suggest a multitude of dissimilar weights for each chemical and geographical variable based on personal geological experience and biases. Giving equal weight to each variable is probably a severe test in cluster analysis. If all major and minor chemical elements are correlated (closed-number data), undoubtedly arguments could be made for data redundancy having biased the resultant clustering.

THE EXACT-TEST APPROACH

Discussion in the previous section could suggest that possible improprieties arise in using cluster analyses for identifying natural groups (suites) among samples, especially those represented by constant-sum chemical data. For this reason, we use Fisher's exact-test (randomization-test) approach (see Snedecor and Cochran, 1980, p. 147) to examine all possible partitionings of the original sample data into groups (potential suites). The method is explained best by a simple example.

Consider the Bemboka (39 samples) and Glenbog (43 samples) suites described by Beams (1980) and Chappell (1984) from Bega Batholith, Lachlan Fold Belt. Although over 30 major- and trace-element analyses are available, Sr and SiO_2 analyses for all 82 samples can be plotted on a (two-dimensional) graph or Harker diagram. With no constraints except that a suite must have at least, say, two samples, all possible pairs of groups (each with two or more samples) that could be drawn from the 82·samples can be identified; each of these p pairs of groups can be tested to determine whether it is better than the particular case represented by the putative Bemboka and Glenbog suites. "Better" in this context presumably means a more well-defined pair of clusters on the graph. If mere discrete groups are sufficient, discriminant analysis could be used to assess the degree to which a linear-discriminant function separates each group; variance of samples about the center of gravity of each group might also be tested. Both Sr and SiO_2 have significant ranges in the alleged suites and the samples tend to be scattered along *linear* belts on the Harker diagrams; on the assumption that this is a characteristic of these suites in general, "better" discrete groups can be identified by the sum of the sum-of-squares reductions associated with least-squares linear approximations to the data in each pair of groups.

The exercise can be repeated using Na_2O and SiO_2. One may seek those q pairs of groups among the total p pairs of groups in which the sum-of-squares reductions (S) for both Sr/SiO_2 and Na_2O/SiO_2 are less than for the alleged Bemboka and Glenbog suites. Alternatively, one can test the plane fitted by least-squares methods in Sr-Na_2O-SiO_2 space for each of the $p - 1$ pairs of groups against that for the putative suites. All other pairs of the 32 variables (e.g., Ga/SiO_2, Pb/Rb) can be tested in like manner. In principle, this technique can be extended by fitting hyperplanes in the multi-dimensional space of several, or all, chemical variables for which analyses are available; various ratios of elements, factor-analysis scores, and other variables could also be used. Additionally, quadratic or larger-degree curves, planes, or hyperplanes could be approximated by least squares instead of linear ones. In principle, all 304 available analyzed samples could be included, rather than only the 82 specimens said to belong to the Bemboka and Glenbog suites; although search could be made for only the p sets of two groups, to search for more than two groups would now be prudent because Beams and Chappell asserted that more than 20 suites actually occur.

Testing whether "better" ways of subdividing the 82 samples assigned to the Bemboka and Glenbog suites exist involves only a simple computer program (see the Appendix). However, with N samples, the number of pairs of groups with at least two samples in each group is $[2^{N-1} - (N + 1)]$; with 81 samples there are almost 2^{81} sets of two groups and applying even a simple linear model to each set would involve astronomical computer time. With the UNIVAC 1100/80 computer available to us, evaluating 20 samples at a time only involved eight minutes computing time. Hence, as an expedient, we examined subsets of samples of the supposed Bemboka and Glenbog suites and of other pairs of supposed suites and obtained convincing results.

Sampling is a major consideration in this procedure. We chose to assume that (1) available chemical analyses reflect the composition of the several kilogram rock samples without error, (2) the rocks collected in the field for chemical analysis constitute a truly random sample of the sampled population, and (3) no problems arise in extrapolating from the sampled populations to the plutons (target populations). Vagaries of where outcrops happen to occur and of varying size and compositional variability of plutons leave something to be desired in this regard, although most plutons in Bega Batholith have much greater homogeneity than granitoids from many other regions (when examined visually at hand-specimen, outcrop, and pluton levels of study). Because the spatial location of postulated suites was unknown until after collection and chemical analysis of samples, only an arbitrary relationship exists between the sampled populations and the possible suites.

Subject to caveats stemming from considerations outlined in the previous paragraph, we assumed for present purposes that the 82 samples for Bemboka and Glenbog suites adequately represent the spatial variability of these suites' outcrops; similarly for the other supposed suites. Selection of 17 to 20 samples from these two suites to afford reasonable computer-processing time can be achieved in numerous ways. The 39 Bemboka rocks come from a single pluton; initially, the eight samples geographically closest to the Glenbog suite outcrops were chosen, along with all samples of either the Nimmitabel or the Towamba Plutons (Glenbog suite), these being plutons close to Bemboka Pluton. This approach yielded clusters of data with small SiO_2 ranges on Sr/SiO_2 and Na_2O/SiO_2 graphs; lines approximated to various pairs of groups drawn from such populations tended to have slopes dissimilar to lines for the whole supposed suites. A different, apparently preferable, method is to choose eight or nine samples from each suite so as to obtain a large variability of SiO_2 values (e.g., the sample with largest, smallest, and median SiO_2, and then successive medians between chosen samples until the selected set is of sufficient size). This method relies heavily on the presumed linear relationship between Sr and SiO_2 (and Na_2O and SiO_2) and gives more precise estimates of the slopes; it was used for the examples described in the next sections.

SOME RESULTS FROM USE OF THE EXACT-TEST APPROACH

Bemboka (39 samples) and Glenbog (43 samples) suites, Bega Batholith

To illustrate the latter method, eight Bemboka and nine Glenbog suite samples were chosen so as to obtain a large variability of SiO_2 values. In addition to Na_2O, Sr, and SiO_2, four other pairs of elements were chosen arbitrarily, namely Rb, Ga, and CaO, and Pb, Cr, and MgO (see Figure 5.2). The 17 samples were then divided into the 65,518 possible pairs of

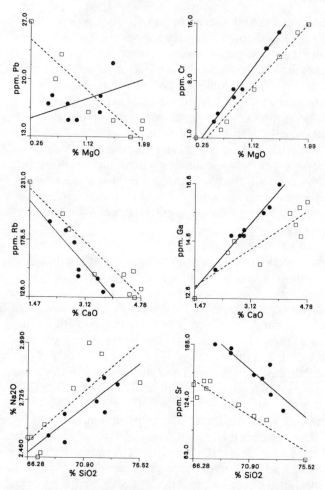

Figure 5.2   Harker-type diagram for data of Chappell and co-workers for the 8 Bemboka (circles, solid lines) and 9 Glenbog (squares, broken lines) suite samples used in this chapter; actual chemical analyses and the linear-regression lines for these data.

groups (all groups containing two or more samples); 26,859 such pairs of groups (almost 41 percent) yielded smaller S values (sums-of-squares reduction when linear approximations are made to data of each of the two groups) with respect to Na_2O/SiO_2 than the lines for supposed Bemboka and Glenbog suite rocks. By contrast, no pair of groups has a smaller S value with respect to Sr/SiO_2 than that for the alleged suites. Hence, considering SiO_2, Na_2O, and Sr simultaneously and assuming a linear relationship, no pair of groups is better than the original suites on the basis of this exact-test experiment based on 17 selected samples.

Using Pb, Cr, and MgO for the same 17 samples gave parallel results; although 257 and 29 pairs are better when Pb/MgO and Cr/MgO are used, respectively, no better pair of groups exists when Cr, Pb, and MgO are used together. However, 2201 and 260 pairs are better when Rb/CaO and Ga/CaO are used, respectively, *and* 24 of the possible 65,518 pairs (3.7 percent) are better than Bemboka and Glenbog suites when Rb, Ga, and CaO are used together.

Why Worry (13 samples) and Candelo (12 samples) suites, Bega Batholith

A similar experiment was run using the same chemical-variable pairs with nine Why Worry and nine Candelo suite samples (Figure 5.3); these suites outcrop in contiguous, but discrete, areas. With Na_2O, Sr, and SiO_2, no better pairs than the alleged suites exist. However, 20,946 and 31,610 of the possible 131,053 pairs of groups are better when Rb/CaO and Ga/CaO are used, respectively, and 5769 pairs of groups are better when Rb, Ga, and CaO are all used together; many samples would be assigned to the other suite. Similarly, 46,282 and 7863 pairs are better when Pb/MgO and Cr/MgO are used, respectively, and 4140 of the possible 131,053 pairs of groups are better when Pb, Cr, and MgO are used together.

Brogo (7 samples) and Moruya (17 samples) suites, Bega and Moruya Batholiths, respectively

Brogo and Moruya suites lie at the eastern margin of Bega Batholith, but the Moruya rocks outcrop more than 60 km north of Brogo suite. Using ten Moruya and seven Brogo samples (Figure 5.4), 292 pairs of groups yield smaller S values for Na_2O/SiO_2, while for Sr/SiO_2, 15 groups yield smaller S values than the original supposed suites. However, only one pair of groups is (marginally) better than the supposed suites when both Sr/SiO_2 and Na_2O/SiO_2 are considered.

Using other pairs of variables leads to similar results. Six and 2476 of the possible 65,518 pairs of groups are better with Rb/CaO and Ga/CaO, and 2922 and 42 pairs with Pb/MgO and Cr/MgO, respectively, but in both cases only one (different) group is better when Rb, Ga, and CaO and Pb, Cr, and MgO are used. The samples reassigned to obtain the marginally better results (than those for the named suites) are:

Variables	Brogo sample	Moruya samples					
$Na_2O/Sr/SiO_2$	LFB70				LBF10,	LFB15	
Rb/Ga/CaO	None	LFB7,		LFB9			
Cr/Pb/MgO	LFB70		LFB8,	LFB9,	LFB10,		LFB16

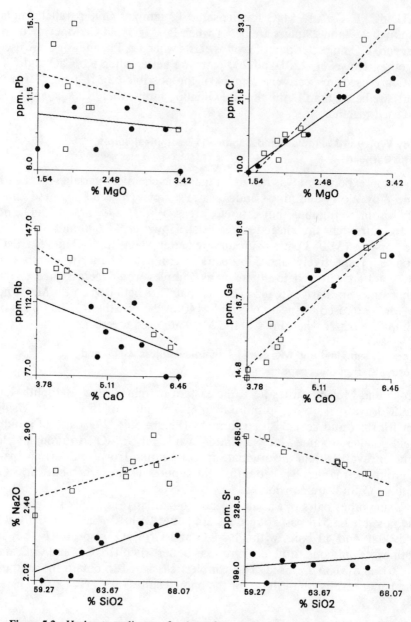

Figure 5.3 Harker-type diagram for data of Chappell and co-workers for the 9 Why Worry (circles, solid lines) and 9 Candelo (squares, broken lines) suite samples used in this chapter; actual chemical analyses and the linear-regression lines for these data.

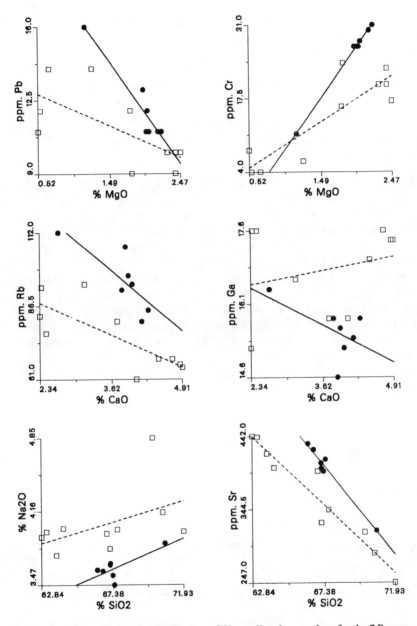

Figure 5.4 Harker-type diagram for data of Chappell and co-workers for the 7 Brogo (circles, solid lines) and 10 Moruya (squares, broken lines) suite samples used in this chapter; actual chemical analyses and the linear-regression lines for these data.

Results for all 231 pairs of the 22 alleged suites cited in Table 5.1 can not be described here. Bemboka/Glenbog, Why Worry/Candelo, and Brogo/Moruya have been chosen as examples because, on Sr/SiO_2 and Na_2O/SiO_2 plots (Figure 5.1), data for these pairs of suites are represented by close, parallel lines; that is, initially, there appears to be a high probability that differentiation between them may be unreal.

Why Worry (13 samples) and Tonghi (15 samples) suites, Bega Batholith

Rocks of the supposed Why Worry and Tonghi suites present a different type of problem. Tonghi suite was mapped (Beams, 1980, Fig. 6) in a 60 km NNE-SSW belt with one small pluton offset a few kilometers northwest of the northern end of the belt; Why Worry suite lies in a small region offset a comparable distance northeast of the northern end of Tonghi suite. Hence, there is no compelling geographic reason for separating the outcrops into two suites. SiO_2 values for rocks of the supposed suites only just overlap and lines for the two suites on Sr/SiO_2 and Na_2O/SiO_2 graphs are essentially continuous, but with slightly different slopes for each suite (Figure 5.1). Use of nine Why Worry and nine Tonghi suite rocks (Figure 5.5) with the exact-test technique suggested that better pairs of groups than the alleged suites exist; either better suites need identification, or all of these samples really belong to one suite, not two. Of 131,053 possible pairs of groups, the number of better pairs of groups are:

With SiO_2/Na_2O alone	57,312
With SiO_2/Sr alone	20
With $SiO_2/Na_2O/Sr$ together	17
With Rb/CaO alone	14,822
With Ga/CaO alone	92,842
With $Rb/Ga/CaO$ together	9,939
With Pb/MgO alone	1,094
With Cr/MgO alone	21
With $Pb/Cr/MgO$ together	5

DISCUSSION

The exact-test (or randomization-test) approach affords a method of quantitatively and objectively testing explicitly whether an independently iden-

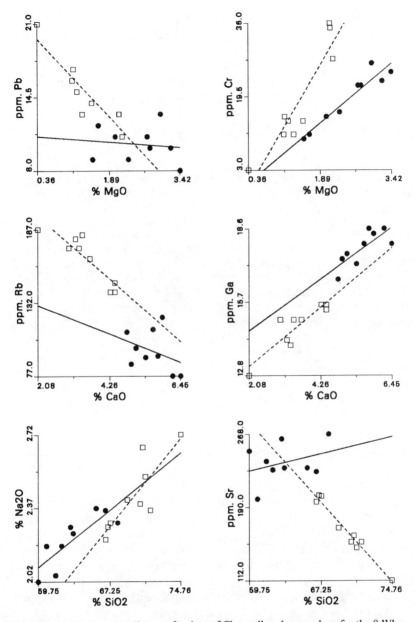

Figure 5.5 Harker-type diagram for data of Chappell and co-workers for the 9 Why Worry (circles, solid lines) and 9 Tonghi (squares, broken lines) suite samples used in this chapter; actual chemical analyses and the linear-regression lines for these data.

tified pair of suites is the "best" pair of sets of rocks that can be drawn from available samples; that is, whether alleged suites represent the "best" natural grouping. Note that, to test objectively whether samples come from a suite, an operational definition of "suite" in terms of the available data must be erected. For this chapter, the distribution of analyses along linear (or higher-degree) belts on plots of two elements is assumed to be an essential characteristic of a suite.

The exact tests described in this chapter provide a rigorous method of testing the reality (validity) of alleged suites. However, for rigorous evaluation, all pairs of 22 suites need testing, suites may contain 40 or more samples that need to be included simultaneously, and it is unrealistic to rely only on SiO_2, Na_2O, and Sr; B. W. Chappell, W. Compston, and A. J. R. White (personal communication, 1981) have asserted that:

> Granitoids within a suite will have many petrographic characteristics in common, but these lack the quantitative aspect necessary to infer derivation from identical sources. Characterisation of a suite can be made using chemical and isotopic criteria and is achieved when all elements vary in concomitant fashion. Suites will normally encompass a range of chemical composition, but the specification of one element must define all others within the limit of sample variation of that suite.

According to Chappell (1984, p. 696)

> Suites are thought to correlate with differences in source rock composition so that different suites result from different sources rather than different processes of crystallization or solidification. Conversely, a single suite is derived from effectively uniform source rocks.

Hence, if Chappell and co-workers' suites are real, elements other than SiO_2, Na_2O, and Sr are expected to permit identification of suites; for example, White and Chappell (1983, Fig. 4) illustrated the use of Ni/SiO_2 to define Jindabyne and Currowong suites.

Such numerical rigorous testing, although theoretically possible, would involve astronomical amounts of computer time. As a result, we have used a heuristic approach compatible with available realistic computer time; we used less than 20 samples at a time, studied only three of four pairs of alleged suites, and relied on only a small selection of the 32 available chemical variables.

This approach failed to indicate that significantly better pairs of groups exist than the alleged suites when attention is focused on Sr, Na_2O, and SiO_2 only, except in the case of Why Worry and Tonghi suites. When Rb, Ga, and CaO, and Pb, Cr, and MgO were used, this heuristic application of the exact-test method suggested that the named suites examined are not justified. However, when Sr, Na_2O, and SiO_2 are the sole bases for decision making, this method lends credence to some of the earlier results based on cluster analyses (Table 5.1), in that many of the previously and independently identified suites in the Lachlan Fold Belt cannot be shown to be inappropriate natural groupings, although an occasional sample may be better placed in a different suite.

Present results lead to the tentative conclusion that either the suites (1) are *not* justified, or (2) are justified, subject to the linear constraints mentioned earlier in this section, *provided* that Sr, Na_2O, and SiO_2 (and possibly other variables, as yet unidentified) only are prescribed as those critical to the definition of the term suite. This conclusion raises important questions about the operational definition of the term suite.

Preliminary work among the 231 pairs of named suites suggests that (1) different elements so far tested in each pair of suites have bimodal and unimodal distributions and (2) that inclusion of one or more bimodal element in an exact-test experiment tends not to permit "better" (smaller S values) groups to be found than the original suites.

A HARDER QUESTION

If the Bega Batholith is characterized by suites, it would be unreasonable not to assume that at least some other batholiths around the world also contain suites. For the Sierra Nevada Range of California (USA), the Okhotsk–Chukchi volcanogenic belt (Soviet Far East), and the Caledonide-Appalachian system (N.W. Europe and E. North America) thousands of chemical analyses are available. Do natural groupings, or suites, occur in the batholiths of these regions?

We have addressed the easier problem: If someone has independently identified suites or other "natural groupings" (probably on the basis of much more geological input than chemical analyses alone), can chemical analyses alone be used to deny or corroborate the reality of suites? Our tentative conclusions (previous section) raise serious complex problems that demand further study; we have initiated further research. More difficult is the question of whether real suites can be identified from chemical analyses available for a batholith (e.g., a batholith for which no geologist has postulated suites). Even if there are no other difficulties with cluster analysis, that method always subdivides a data set into clusters; for a batholith, such clusters might, or might not, have the nature of suites. We are addressing this more general topic and plan to make this the subject of a subsequent paper.*

*Some of these issues were addressed as an important part of a paper completed after this chapter was finalized (Whitten et al., 1987).

APPENDIX

The exact-test (randomization-test) approach used in this chapter involves two steps. Assume that n rock samples comprise specimens representing two previously and independently identified suites A and B. Suppose that variables Sr and SiO_2 are available for all samples.

In the first step, the n samples are partitioned into all possible pairs of groups (one of which happens to comprise suites A and B). Suppose that suites A and B comprise four samples $\{a, b, c, d\}$; possible partitionings are $\{a\}$ with $\{b, c, d\}$, or $\{a, c\}$ with $\{b, d\}$, etc. For efficient programming Gray Code (e.g., Ralston, 1983) is used. To construct a Gray Code matrix, the first $2^{(I-1)}$ elements in the Ith column have the same sign, and, from the $\{2^{(I-1)} + 1\}$th element downward in the Ith column, each of 2^I elements has the opposite sign; if additional elements down the Ith column are needed, the sequence is repeated. For $\{a, b, c, d\}$, the corresponding Gray Code set is:

$$
\begin{array}{rrrr}
[\,-1 & -1 & -1 & -1\,] \\
[\,1 & -1 & -1 & -1\,] \\
[\,1 & 1 & -1 & -1\,] \\
[\,-1 & 1 & -1 & -1\,] \\
[\,-1 & 1 & 1 & -1\,] \\
[\,1 & 1 & 1 & -1\,] \\
[\,1 & -1 & 1 & -1\,] \\
[\,-1 & -1 & 1 & -1\,] \\
[\,-1 & -1 & 1 & 1\,] \\
[\,1 & -1 & 1 & 1\,] \\
[\,1 & 1 & 1 & 1\,] \\
[\,-1 & 1 & 1 & 1\,] \\
[\,-1 & 1 & -1 & 1\,] \\
[\,1 & 1 & -1 & 1\,] \\
[\,1 & -1 & -1 & 1\,] \\
[\,-1 & -1 & -1 & 1\,]
\end{array}
$$

Each of the 16 rows represents one partitioning; with 1 the sample is assigned to group I and with -1 to group II. Because the order of the groups is irrelevant, the first eight rows of binary numbers represent all possible partitionings of four samples. For large numbers of samples, the appropriate larger matrix of binary numbers is rapidly constructed. The advantage of the Gray Code is that only one element is changed in moving to each successive row (partitioning).

In the second step, each pair of groups (developed by the partitioning) is considered separately. For each group of samples, a linear equation is approximated to the Sr and SiO_2 values by least-squares methods and the sum (S) of the sum-of-squares reductions for both of the two lines (for each pair of groups) is compared with the corresponding S value for the suites A and B. The "better" the pair of groups, the smaller the S value because the chemical values (variables) are more satisfactorily represented by the two approximated lines.

The standard linear least-squares method is used to approximate simultaneously the points in Sr-SiO$_2$ space for each successive partitioning; in practice, S is determined for each pair of groups (without actually calculating the parameters of the two lines), as follows:

$$S = \sum_{J=1}^{2} \{P_J Q_J - R_J^2\} \cdot \{n_J(n_J - 2)P_J\}^{-1}$$

where n_J signifies the number of samples in the jth of the two groups, the Ith of the n_J samples has the chemical composition (x_I, y_I), in this case, the Sr and SiO$_2$ values, and

$$P_J = n_J \Sigma x_I^2 - (\Sigma x_I)^2$$

$$Q_J = n_J \Sigma y_I^2 - (\Sigma y_I)^2$$

and

$$R_J = n_J \Sigma x_I y_I - (\Sigma x_I)(\Sigma y_I)$$

Because groups in successive partitionings have only one different sample with the Gray Code, successive S values are obtained with the following formulae:

$$\text{new } R_1 = (n_1 - 1)(\Sigma XY - x_I y_I) - (\Sigma X - x_I)(\Sigma Y - y_I)$$

and

$$\text{new } R_2 = (n_2 + 1)(\Sigma XY + x_I y_I) - (\Sigma X + x_I)(\Sigma Y + y_I)$$

where ΣX, ΣY, and ΣXY are the last determined Σx_I, Σy_I, and $\Sigma x_I y_I$ values, and x_I and y_I are the values of the current Ith sample moved from group $J = 1$ to group $J = 2$.

When necessary, the two straight lines $y_J = a_J + b_J x_J$, $(J = 1,2)$ are obtained from:

$$b_J = R_J \cdot P_J^{-1}$$

and

$$a_J = \bar{y}_I - b_J \bar{x}_I$$

where $\bar{x}_I$ is the mean value of x_I in group J and $\bar{y}_I$ is the mean value of y_I in group J.

Use of these formulae on a Univac 1100/80 computer takes 56 seconds,

2, 4, and 8 minutes for 17, 18, 19, and 20 samples, respectively, whereas use of common binary-code systems would involve some 10 minutes when only 17 samples are processed. (Because major elements in weight percent and trace elements in parts per million are involved, the analytical data for each variable were always normalized to have a range from -1 to $+1$ prior to processing).

ACKNOWLEDGMENTS

Special thanks are due to Dr. B. W. Chappell, who enabled one of us (E.H.T.W.) to gain first-hand knowledge of Lachlan Fold Belt in 1981 during sabbatical leave at Australian National University. Dr. Chappell encouraged investigation of the statistical characteristics and significance of abundant chemical analyses for Bega Batholith produced in his laboratory. To this end, complete analytical data for 1000 samples were made available on tape; these data are the bases for White and Chappell (1983) and Chappell (1984), and many were listed by Beams (1980), but many are unpublished (e.g., Beams, Chappell, and White, in preparation). Computing facilities at Australian National University and Michigan Technological University, and a 1980–1981 travel grant (to E.H.T.W.) from the American Philosophical Society are acknowledged.

REFERENCES

Aitchison, J., 1984, The statistical analysis of geochemical compositions: *Jour. Int'l. Assoc. Mathematical Geology*, v. 16, p. 531–564.

Bayly, M. B., 1968, *Introduction to petrology:* Prentice-Hall, Englewood Cliffs, N.J., 371 p.

Beams, S. D., 1980, Magmatic evolution of the southwest Lachlan Fold Belt, Australia: Ph.D. Thesis, La Trobe University.

Chappell, B. W., 1978, Granitoids from the Moonbi district, New England Batholith, Eastern Australia: *Geol. Soc. Australia Jour.*, v. 25, p. 267–283.

Chappell, B. W., 1984, Source rocks of I- and S-type granites in the Lachlan Fold Belt, southeastern Australia: *Phil. Trans. Royal Soc. Lond., A*, v. 310, p. 693–707.

Chayes, F., 1971, *Ratio correlation: a manual for students of petrology and geochemistry:* Univ. Chicago Press, Chicago, 99 p.

Chayes, F., and Kruskal, W., 1966, An approximate test for correlations between proportions: *Jour. Geol.*, v. 74, p. 692–702.

Dixon, W. J. (chief ed.), 1981, *BMDP Statistical Software:* Univ. of California Press, Berkeley, 726 p.

Griffin, T. J., White, A. J. R., and Chappell, B. W., 1977, The Moruya Batholith and a comparison of the chemistry of the Moruya and Jindabyne Suites: *Geol. Soc. Australia Jour.*, v. 25, p. 235–247.

Miesch, A. T., 1969, The constant sum problem in geochemistry: *in* Merriam, D. F. (ed.), *Computer applications in the earth sciences:* Plenum Press, New York, p. 161–176.

Pitcher, W. S., 1982, Granite type and tectonic environment: *in* Hsu, K. J. (ed.), *Mountain Building Processes:* Academic Press, New York, p. 19–40.

Pitcher, W. S., 1983, Granite: typology, geological environment and melting relationships: *in* Atherton, M. P., and Gribble, C. D. (eds.), *Migmatites, melting and metamorphism:* Shiva Pub. Ltd., Nantwich, p. 277–85.

Ralston, A. (ed.), 1983, *Encyclopedia of computer science and engineering,* 2nd ed.: Van Nostrand Reinhold, New York, 1664 p.

Snedecor, G. W., and Cochran, W. G., 1980, *Statistical methods,* 7th ed.: Iowa State Univ. Press, Ames, 608 p.

White, A. J. R., and Chappell, B. W., 1983, Granitoid types and their distribution in the Lachlan Fold Belt, southeastern Australia: *Mem. Geol. Soc. Amer.* 159, p. 21–34.

Whitten, E. H. T., 1985, Suites within a granitoid batholith: A quantitative justification based on the Lachlan Fold Belt, S.E. Australia: *Geol. Zborn., Geol. Carpath. (Bratislava),* v. 36, p. 191–199.

Whitten, E. H. T., Bornhorst, T. J., Li, G., Hicks, D. L., and Beckwith, J. P., 1987, Suites, subdivision of batholiths, and igneous-rock classification: Geological and mathematical conceptualization: *Amer. Journ. Sci.,* v. 287, no. 4, p. 332–352.

Whitten, E. H. T., and Chappell, B. W., 1984, Suites within a granitoid batholith: A quantitative justification based on the Lachlan Fold Belt, SE. Australia: *Abstr. XXVII Internat. Geol. Congr. (Moscow),* v. IV, p. 489.

6

Misuses of Linear Regression in Earth Sciences

C. John Mann

Students beginning statistics learn about linear models almost as soon as they learn mean and standard deviation. Next to these two most commonly used statistical indices, linear regression and its cousin, analysis of variance, are probably the most commonly used statistical techniques in science. Regression has been called the most important tool of the applied statistician (Wonnacott and Wonnacott 1981). Unfortunately, linear regression all too often is used incorrectly.

Incorrect uses can be attributed probably to three primary causes. First, widespread availability of computers and programs to perform regression analysis makes its use easy for anyone, whether knowledgeable in statistics or not. It is often a nice way to demonstrate relationships, either rightly or wrongly. These are errors of inappropriate use and could be called errors due to ignorance. Second, linear regression commonly is applied to data with no concern as to whether assumptions inherent to the technique are met. In the past, these assumptions probably have been ignored because suitable tests to demonstrate validity that a specific data set meets assumptions were either nonexistent, were not widely known, or were computationally difficult by hand. Rarely in papers utilizing linear regression do nonmathematicians demonstrate that their data meet all criteria demanded by theory. These shortcomings could be called errors due to laziness. Third, even when all data, calculations, and assumptions lead to an impeccable mathematical analysis, some results of linear regression are extended incorrectly beyond the data base or misinterpreted to draw unwarranted geological conclusions. These could be called errors due to poor judgment.

During the past 25 years, however, applied statisticians have directed considerable attention to these problems and have developed numerous ways in which to check assumptions or modify data, so that applications of linear regression to real data and real problems more closely meet theoretical requirements. The theory has not changed. All that has changed is our understanding and ability to handle natural data more realistically now.

Many errors, misuses, and abuses of statistics in earth sciences arise, in part, because relatively few geologists have had adequate training in statistical mathematics, yet many are using these techniques. Geological data are complex, noisy, and commonly inadequate to draw clear and accurate conclusions about nature. Consequently statistical theory offers to us a reasonably objective and reproducible basis for handling our naturally variable and limited information. Its use should be more widespread than it already is.

Remarks and observations in this chapter are dominantly qualitative to be readable and to appeal to the greatest number of earth scientists possible. Interested readers will find quantitative and mathematically rigorous treatments of various topics covered in references throughout the chapter. Examples taken from recent geological literature to serve as vehicles for demonstrating a few enlightening techniques and their easy application should not be construed as condemnation of the authors cited because the use of linear regression in earth sciences should be encouraged, not discouraged. Blame for inappropriate applications in geological literature must rest rather with older geoscience sages who are journal editors and reviewers of the papers published but who do not fulfill responsibilities inherent in their positions and do not counsel authors wisely during manuscript preparation stages. Referees rarely check equations, numbers, and numerical results against data.

Correct linear regression analyses are relatively easy. All you have to do is use common sense and understand clearly a few significant points. Don't use statistics, mathematics, or a computer if you don't understand them even though it might be easiest, currently fashionable, or professionally smart to do so; either learn the methods thoroughly first or get some wise, competent counseling during its use. Do this and you won't go too far wrong.

LINEAR REGRESSION MODELS AND THEIR BASIS

The linear regression model commonly is written as

$$Y = \alpha + \beta X + \delta$$

where X is an independent, normally distributed variable used to predict Y, another independent, normally distributed random variable, α is the Y-axis intercept of the linear prediction equation when $X = 0$, β is the slope of the predictor line, and δ is a random (or stochastic) disturbance term. Commonly, δ is called a random error or residual term in the equation; it is the amount of variation in Y for which the linear relationship does not account and it may represent observation error as well as the effects of unspecified additional variables and a natural random element (noise) or variation in the X-Y relationship. More specifically,

$$y_i = \alpha + \beta(x_i - \bar{x}) + \delta_i \quad \text{and} \quad \hat{y}_i = a + b(x_i - \bar{x})$$

where

$$a = \bar{y} - b\bar{x} \quad \text{and} \quad b = \Sigma(x_i - \bar{x})(y_i - \bar{y})/\Sigma(x_i - \bar{x})^2$$

Variable Y has been called the dependent variable, response variable, or predictand for purposes of regression analysis, whereas variable X has been referred to as predictor variable, regressor variable, independent variable, or factor variable (Wonnacott and Wonnacott, 1981). Originally, regression analysis referred only to both variables being observed randomly (Snedecor, 1956) (Figure 6.1, upper) but now it generally is accepted that one variable may be either "fixed" (nonrandom) as well as random (Graybill, 1961, 1976; Draper and Smith, 1981) (Figure 6.1, lower).

Linear regression apparently originated about 1794 to 1805 through work of Carl Friedrich Gauss and Adrien Marie Legendre (Seal, 1967) although some dispute exists about which one of them (Plackett, 1972; Stigler, 1981) first used least squares as a method for fitting linear equations to data. Linear regression subsequently was placed on a more firm theoretical basis by Karl Pearson in 1896 (Seal, 1967) who linked it with the multivariate normal probability density distribution that had evolved from 1846 to 1892.

Thus, several properties are assumed to be true whenever linear regression is applied to data. Theoretically, the linear regression model is valid *only* when all assumptions are binding (Scheffé, 1959). These assumptions are:

1. X and Y are measured without error.
2. X and Y are normally (Gaussian) distributed, independent variables.
3. A linear relation exists between X and Y (Figure 6.2).
4. X values are linearly independent.
5. Errors (δ) of prediction are normally distributed with a mean of zero (Figure 6.2).
6. Errors (δ) have a variance that is constant (homoscedasticity) (Figure 6.2).
7. Errors (δ) are serially independent (not autocorrelated).

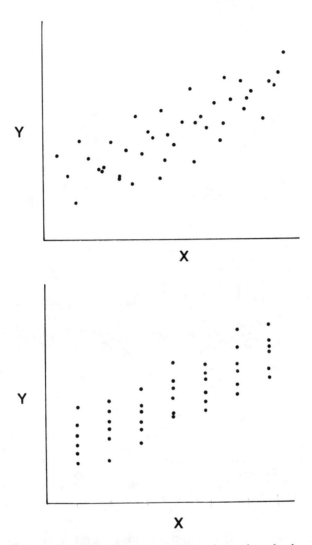

Figure 6.1 When both X and Y are observed randomly (upper), pairs of observations will define a true bivariate normal distribution if a linear relationship exists and X and Y are distributed normally. However, when one variate is fixed (nonrandom) during observations (such as X, lower), pairs of X-Y values will not define a bivariate normal distribution even though X and Y are sampled, in fact, from a bivariate normal distribution.

Clearly, as you know, no one can measure X and Y under any conditions without error; therefore, all applications of linear regression are incorrect technically. However, many mathematicians have shown that this assumption is not crucial for many types of applications and can be relaxed safely (Eisenhart, 1939; Wald, 1940; Winsor, 1946; Bartlett, 1949; Berkson,

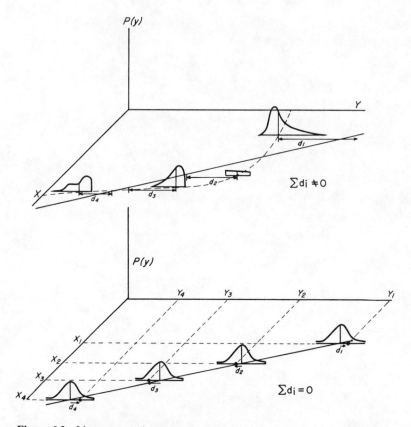

Figure 6.2 Linear regression analysis assumes that a linear relationship exists between X and Y (bottom) and that disturbance terms (d_i, bottom) are normally distributed with means of zero and equal variance. If the relationship, in reality, is nonlinear (top), has deviations with means that are non-zero, has variances that are unequal, or variances that are not normally distributed (top), the linear regression model is inappropriate and should not be used.

1950; Madansky, 1959; Gunst and Mason, 1980). For example, measurement error in X and Y can be ignored, as well as all assumptions (Gunst and Mason, 1980), if the sole purpose of the regression analysis is to use X to predict values of Y (Lindley, 1947; Johnston, 1963; Mark and Church, 1977). Lindley (1947) argues, for this reason, that linear regression analysis should be restricted entirely to purposes of prediction (see also Mark and Church, 1977) and rather that linear functional analysis, a related statistical technique, be used to determine coefficients for describing a linear relationship between two independent variables measured with error in natural sciences.

Commonly, only X is considered to be observed without error of measurement whereas errors of Y measurement are considered to be included in δ. However, measurement errors in X will typically cause least-square estimators of regression coefficients to be biased (Berkson, 1950; Swindel and Bower, 1972); least squares is valid only if random variation in predictor variable is so small, as compared to predictor variable range, that it can be ignored (Draper and Smith, 1981). Functional analysis contrarily will provide unbiased estimators for coefficients even if X and Y do have measurement errors (Lindley, 1947). This has led some to say (Gunst and Mason, 1980) that theoretically all regression models are only approximations.

Twenty five years ago neither the statistical art nor use of computers was sufficient to test these assumptions adequately by nonmathematicians (Hocking, 1983). As a result, both statistical courses and textbooks generally ignored practical aspects of assumption testing when regression analyses were applied to natural data. Tacitly, users were encouraged to ignore the problem of determining if assumptions fundamental to linear regression models were valid. This is no longer true.

Various tests and solutions (Table 6.1) are available now that can be applied by nonmathematicans to make certain that their linear regression analysis meet these assumptions or, if not, whether we should be concerned with the results. Some tests are even starting to be covered in recent geoscience statistical texts (Davis, 1973).

The assumptions that most commonly are violated in applied regression analysis are those of linearity and homoscedasticity (Chatterjee and Price, 1977). Both may be checked easily by examining residuals (difference between observed values and predicted values). Therefore, residual plots should be made routinely against $\dot{X}$ and $\hat{Y}$ (predicted Y) whenever regression analysis is performed. Residual plots are extremely valuable; they can be used profitably for recognizing patterns in the data. They help to reveal trends, extreme values (outliers), potential problems in the analysis, violations of linear regression assumptions, and provide indications of predictor variable importance. Residual analysis has even been called the "most important task in any regression analysis" (Gunst and Mason, 1980). Nonetheless, one can rarely be certain, even using residual analysis, that a linear regression model has been correctly specified.

A growing concern and recognition by applied statisticians over failure of meeting assumptions inherent in linear regression analysis has led to new ways of performing these tasks that avoid, reduce, or relax these criteria. Linear functional analysis, mentioned previously, was one of the earliest techniques. A nonparametric scheme has been proposed (Hogg and Randles, 1975) that does not require an assumption of normality or symmetry in the independent variables. Winsorized regression (Chen and Dixon, 1972; Yale and Forsyth, 1976) has been proposed to reduce effects of "contamination" of a sampled population by diminishing effects of "outliers." Contamination is used here in the sense that some observations

TABLE 6.1

Assumption	Tests	Solutions
Each X and Y observed without measurement error. (Often relaxed so that only X is observed without error and δ expanded to include errors of Y measurement.)	Difficult to test, thus rarely examined. Check for operator errors (Chatterjee and Price, 1977). Multiple data set collection. Assume errors are independent and normally distributed, then use estimates of error variances as weights in modified least-squares model (Poole and O'Farrell, 1971).	Measure data more accurately. Measure data more carefully. Limit analysis to theoretically valid portion of data collected.
Normality—For "random" X model, both conditional and marginal distributions of X and Y must be normal. For "nonrandom" X model, conditional distribution of disturbance term must be normal, implying that Y's are normally distributed.	Test for normality (Dyer, 1974; Oja, 1983; Olsson, 1979; Pierce and Gray, 1982; Shapiro and Wilk, 1965; Tartar and Kowalski, 1972). Plot d_i against expected order statistic $$\Phi^{-1}\left(\frac{i - \frac{1}{2}}{n}\right) \text{ from}$$ standard normal distribution (Draper and Smith, 1981; Seber, 1977). Test for outliers (Wooding, 1969). Shapiro–Wilk (1965) test. Anderson–Darling statistic (Hawkins, 1981). Geary test (DiAgostino and Rosman, 1974; Shapiro, Wilk, and Chen, 1968).	Transform data, if possible (Emerson and Stoto, 1983; Pericchi, 1981; Tartar and Kowalski, 1972).
Multicollinearity—Independent variables X and Y are linearly independent of each other.	Revealed, generally, by large standard errors (Poole and O'Farrell, 1971). Confluence analysis (bunch-map analysis) (Johnston, 1963).	Limit analysis to pairs of variables which do not show evidence of multipollinearity.

TABLE 6.1 (*Continued*)

Assumption	Tests	Solutions
	Small characteristic roots of correlation matrix (Chatterjee and Price, 1977). Kendall correlation coefficient (Wolfe, 1977).	
Linearity—Actual relationship between X and Y is a linear one.	Generalized likelihood ratio test (Green, 1971). Check for outliers (Chatterjee and Price, 1977; Seber, 1977). Fit polynomial curve to data and test coefficients for significant departure from zero (Poole and O'Farrell, 1971; Seber, 1977). Stratify data for X values and fit regression to each subset; test for significant differences in slopes and intercepts (Green, 1971; Snee, 1977; Tsutakawa and Hewett, 1978). Test residuals, in order of X, for randomess (Chatterjee and Price, 1977; Seber, 1977). Determine confidence interval and compare with data falling outside.	Transform data, if possible, to make a linear relationship (Chatterjee and Price, 1977; Emerson and Stoto, 1983). Restrict analysis to smaller subsets of data (Chatterjee and Price, 1977).
Distribution of disturbance term (d_i or δ_i) is normal with a mean = 0.	Residual analysis patterns. (Goodall, 1983; Seber, 1977). Plot residuals against $\hat{y}_i$. Plot residuals against X_i. Test for outliers (Wooding, 1969).	Check data for errors (Chatterjee and Price, 1977). Use different predictor model (Draper and Smith, 1981). Eliminate some data (Chatterjee and Price, 1977). Collect more data (Draper and Smith, 1981).

TABLE 6.1 (*Continued*)

Assumption	Tests	Solutions
Homoscedasity— Conditional distribution of d_i or δ_i has a constant variance.	Recursive residuals (Harvey and Phillips, 1974). Bickel's test (Bickel, 1978). Hammerstrom's (1981) test Cook–Weisberg's (1983) test Residual analysis (Anscombe and Tukey, 1963; Chatterjee and Price, 1977; Cox and Snell, 1968; Draper and Smith, 1981; Seber, 1977). Plot residuals against x_i. Plot residuals against $\hat{y}_i$. Plot squared residuals against $\hat{y}_i$ (Gunst and Mason, 1980). BLUS test (Harvey and Phillips, 1974). Goldfield and Quandt test (Harvey and Phillips, 1974).	Check data for errors (Chatterjee and Price, 1977). Use different predictor model (Draper and Smith, 1981). Eliminate some data (Chatterjee and Price, 1977). Collect more data (Draper and Smith, 1981).
Disturbance terms (δ_i) are serially independent (no autocorrelation).	Run tests (Miller and Kahn, 1962; Olmstead, 1958). Durbin–Watson statistic (continuity ratio) (Blattberg, 1973; Chatterjee and Price, 1977; Durbin and Watson, 1950, 1951, 1971; Harrison, 1975; Schmidt and Guilkey, 1975). Plot consecutive pairs of time-ordered residuals (Miller and Kahn, 1962; Seber, 1977). Geary tests (Harrison, 1975; Schmidt and Guilkey, 1975).	Use different predictor model (Draper and Smith, 1981).

may have been gathered from a population other than that targeted in the regression analysis.

MISUSES OF LINEAR REGRESSION IN EARTH SCIENCES

Correlation coefficient

Significance. The correlation coefficient is a measure of the linear association between X and Y or an indication of how well X explains, or predicts, Y. If the linear regression equation derived from the analysis predicts exactly each observed Y value, all disturbance terms (δ) will equal zero and the regression equation will have a coefficient of correlation of 1.0, either $+1.0$ or -1.0. Negative values indicate merely that the relationship is inverse, one variable is increasing exactly as the other variable decreases or slope of the graphed relationship is negative, whereas positive values for correlation coefficients are indicative of a direct relationship or an equation with a positive slope. If the coefficient of correlation is zero, the variables are considered to be uncorrelated or not linearly associated; this does not mean, however, that the variables are statistically independent (Draper and Smith, 1981). The coefficient of correlation is calculated by:

$$r_{xy} = \frac{\Sigma(x_i - \bar{x})(y_i - \bar{y})}{[\Sigma(x_i - \bar{x})^2]^{1/2}[\Sigma(y_i - \bar{y})^2]^{1/2}}$$

Correlation coefficients have only relative values. No absolute values of superiority or inferiority can be associated with coefficients obtained from different data sets. For example, a correlation coefficient of 0.90 from one set of data cannot be compared indiscriminately with a coefficient of 0.80 from another set of data and concluded to be superior. One can compare a coefficient for X vs. Y with a coefficient for Z vs. Y derived from the same data set and say that, relatively, X is a better or poorer predictor of Y than Z. Values of the correlation coefficient still remain relative, however, and a coefficient of 0.80 is not twice as good as 0.40 when comparing pairs of variables from the same data set.

The reason for this, in part, is because invariably a random component will be present in data. Rarely will correlation coefficients ever be exactly 1.0 or 0.0. An example of the former is seen in endless numbers of laboratory experiments conducted on known, presumably precisely predictable, linear relationships existing in nature that nonetheless only approach a precisely linear relationship over numerous experiments. In the latter case, we invariably calculate a nonzero coefficient of correlation using random numbers to represent two variables; they will average to zero only over numerous experiments.

What constitutes a significant correlation coefficient? The decision as to what is significant is entirely subjective. What may be significant in one case or from one data set or from an analysis between two given variables may not be significant in another instance. Certainly a correlation coefficient becomes more attractive as it approaches ± 1.0 and less attractive as it approaches 0.0 but to say that it must be greater than z to be significant or excellent is subjective judgment. An excellent illustration of this is cited by Stransky (1984) who reports a correlation coefficient of 0.9924 for rock mechanics data and a linear regression equation that clearly cannot be correct because it predicts a condition that cannot occur in nature.

Similarly, coefficients of correlation only indicate the degree of linear association two variables have mathematically. It does not mean, and cannot be construed to mean, any causal relationship in nature or even that the variables have a physical association of any type. Neither does the correlation coefficient indicate anything about the statistical significance of the linear relationship; it is only a measure of association mathematically in the apparent linear relationship.

A correlation coefficient theoretically is valid only when X and Y form a bivariate normal distribution (assumptions 2 and 3) (see Seber, 1977). When a linear regression is computed for data obtained under controlled conditions, that is, one variable is fixed (Figure 6.1, lower), one may argue (Berkson, 1950) that the correlation coefficient is meaningless because data were not obtained from a definable and significant population. Thus, under these conditions a correlation coefficient has absolutely no scientific meaning.

Spurious self-correlation.   Karl Pearson (1897) noted early in the development of statistics for the solution of scientific problems that a false correlation of variables X and Y could arise whenever they are not truly independent of one another—an assumption of nonmulticollinearity in linear regression analysis. Spurious self-correlation may occur in many ways, but most commonly in natural phenomena it is a result of either X or Y being dependent upon a third variable which is common to a part or all of the other variable. Similarly, X and Y may have a common component. For example, Pearson recognized spurious self-correlations originally as he was comparing bone measurements from one animal; each bone was different but all had the commonality of one host organism. Thus, X and Y were neither random nor independent.

Felix Chayes (1949) noted that spurious self-correlations are found in much geological data because we often are dealing with closed systems with limited numbers of variables; determination of, or fixing, one variable strongly or totally constrains the remaining variables. Similarly, we commonly treat geological data as percentages, which is equivalent to a closed system. He demonstrated that these correlations were false, using petrographic data in several papers (1948, 1949, 1960, 1962, 1971; Chayes and Kruskal, 1966). Spurious self-correlation in geosciences also have been noted by Vistelius and Sarmanov (1961), Benson (1965), Yalin and Kamphuis (1971), and Kenny (1982).

Spurious self-correlation may be demonstrated merely by using random numbers. If X and Y are paired random numbers, a plot of one against the other (Figure 6.3, upper left) will show a random distribution with a randomly oriented linear regression having a correlation coefficient near zero. However, if we plot $X + Y$ (Figure 6.3, upper right), or XY (Figure 6.3, lower left), or Y/X (Figure 6.3, lower right) against X, these random numbers will no longer be randomly distributed, and a regression analysis will indicate correlation coefficients that approach or may even surpass 0.5!

Spurious self-correlations may arise also if disturbance terms (δ) are not serially independent; that is, some autocorrelation exists (Champernowne, 1960; Granger and Newbold, 1974; Pesaran and Slater, 1980). This is an important consequence of using ordinary least squares estimates to obtain the regression equation.

Linear or multilinear regression analysis may be used to explore complex geologic data in attempts to understand better relationships existing in nature. One must be careful, however, in drawing conclusions from correlation coefficients in these instances because spurious correlations might be present. They may enter either because of the way we handle observed data (Figure 6.4) or because we fail to recognize variables that are collinear.

False correlation coefficients arise because variables are not linearly independent; that is, collinearity or multicollinearity exists or variables represent a closed number system.

Collinearity

An assumption that X and Y (Table 6.1) are independent variables for linear regression analysis is often difficult to prove in natural sciences data. Few variables in nature are truly independent physically (or geologically). Time is clearly one that is independent (physically, conceptually, and statistically); but beyond time, few others safely can be stated to be truly independent in all situations. We may be able to include others under local situations and conditions with which we are dealing, but these will not be mathematically independent invariably.

Methods to detect collinearity in regression have been discussed by many workers (Farrar and Glauber, 1967; Leamer, 1973; Kumar, 1975; Wichers, 1975; O'Hagan and McCabe, 1975; Willan and Watts, 1978; Park, 1981). Correlation matrices for predictor variables (C) and for normalized parameter estimators are helpful in revealing sources of collinearity whereas ($C^{1/2}$) provides a measure of overall efficiency (Willan and Watts, 1978).

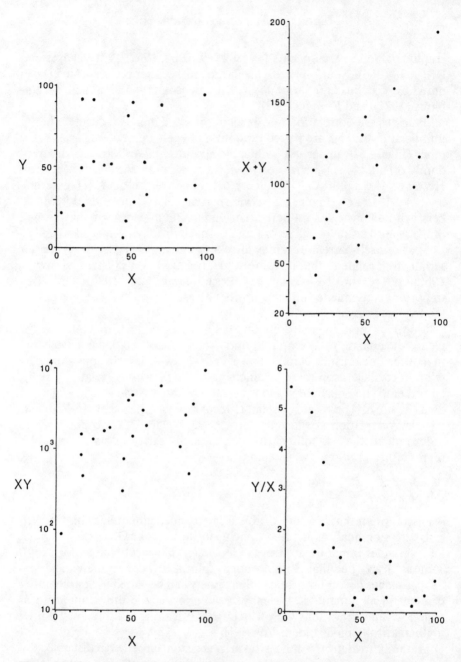

Figure 6.3 Four plots of 20 paired random numbers X and Y. When plotted against each other (upper left), the distribution is random as expected. A linear regression analysis for these data is meaningless; the regression line calculated would be randomly oriented with a correlation coefficient near zero. However, plots of the sum (upper right), the product (lower left), or quotient (lower right) of these random numbers against either X or Y will not be random. Definite linear or curvilinear relationships will exist. A spurious or nonsense self-correlation will have been generated with correlation coefficients near 0.50 for each regression line fitted.

	Q†	w†	d†	w/d†	ū	S†	D_k4†	d/D_k4†	ū/u*	ū/u* bed	ū/u* bk	A/λ†
Q†	1.00	.92	.79	.25	.87	.48	.86	-.35	-.26	-.39	.39	.69
w†		1.00	.61	.57	.75	.48	.79	-.47	-.30	-.43	.41	.65
d†			1.00	-.33	.40	.09	.60	.14	-.24	-.26	-.05	.38
w/d†				1.00	.40	.47	.31	-.15	-0.7	-.24	.24	-.78
ū					1.00	.69	.82	-.59	-.21	-.17	.40	.80
S†						1.00	.82	-.80	-.72	-.82	-.19	.85
D_k4†							1.00	-.70	-.48	-.60	.26	.75
d/D_k4†								1.00	.41	.53	-.18	-.63
ū/u*									1.00	.97	.68	-.13
ū/u* bed										1.00	.56	-.25
ū/u* bk											1.00	.53
A/λ†												1.00

† indicates log values.

Figure 6.4 Correlation coefficients from regression analysis used in exploratory studies (Prestegaard, 1983) must be used with extreme caution. Spurious self-correlation may be introduced because of the manner in which data are handled (w vs. w/d: Q vs. w; Q vs. d; w/d vs. d/D: ū vs. ū/u: ū vs. ū/u*bk) or because of inherent collinearity of variables (Q vs. w; Q vs. d; Q vs. D_{84}; Q vs. A/λ).

Normally, X and Y will be recognized as being mathematically and physically dependent, at least in part, upon a third variable, or even a fourth or more variables, which control both X and Y indirectly. Hence, the correct question in regression analysis of geoscience data normally is not whether X and Y are mathematically independent variables because they rarely will be but, rather, are they sufficiently independent not to invalidate results of regression analysis? If both X and Y knowingly are controlled geologically by a third variable, one can safely conclude in most cases that they are dependent from both a mathematical analysis standpoint and a geological standpoint.

Unfortunately, the answer commonly will lie in the gray zone beyond this clear-cut, easy decision. No definitive criteria have been established to aid in these decisions; only personal judgments can be made. But all too often, no consideration is given and no judgment at all is made concerning this question in regression modeling of earth science data.

Variable selection

Regression analysis may be used for many different purposes. Those variables that may be best for one purpose or in one situation may not be best for another purpose or in another situation; therefore, no "best" set of variables normally will exist for all regression analyses. Rather, the purpose for which a regression analysis is being performed should govern selection of variables that most successfully accomplish this purpose (Chatterjee and Price, 1977).

Because more than one pair of variables may perform the same purpose but with varying degrees of success, consideration of the merits and disadvantages of all possible pairs of variables should be noted by users of regression analysis. This is far superior to merely assuming or arbitrarily selecting one set or generating a single "best" set subjectively. Examining various possible pairs invariably reveals some of the structure of data being used and helps us to understand better those natural relationships and processes involved. Complex situations can be understood more effectively by an iterative approach (Box and Hill, 1967) in which the best mathematical model and variable selection are derived through a series of repeated calculations. This increases the efficiency of experimentation and reduces the numbers of contradictory conclusions.

Much literature delves into this topic in greater detail (Gorman and Toman, 1966; Box and Hill, 1967; Hocking, 1976; Chatterjee and Price,1977; Tsutakawa and Hewitt, 1978). In general, if no geological basis is known for accepting or rejecting potential predictor variables, that variable having the largest, absolute value correlation coefficient with the predicted variable should be selected (Hotelling, 1940).

More important, however, use of $Y = \alpha + \beta X + \delta$ postulates that no other intervening predictor variables exist between X and Y and that X is a unique predictor variable (Gunst and Mason, 1980). All too often, at least in geosciences, if not all sciences, this is not true. We commonly use X merely because we have it, or because it is easy to relate to Y, or because we have more X-Y data than Z-Y data.

Variable selection is an important factor in what John Tukey (1960) has referred to as "statistical vs. experimenter's" conclusions. All mathematical analyses in applied science necessarily result in two significantly different conclusions. "Statistical" conclusions apply only to conditions of the experiment, data, and analysis. These conclusions will always be stronger than the "experimenter's" conclusions because the latter must include all factors entering statistical conclusions as well as other nonstatistical factors. Statistical conclusions may be immanently correct even though inappropriate variables have been included in an experiment, important ones have been omitted, measurements were made incorrectly, or were made with consistent bias, whereas experiment's conclusions, in these instances, would be totally incorrect.

Regression of Y on X

Linear regressions may be handled many different ways (Miller and Kahn, 1962; Jones, 1979; see Troutman and Williams, Chapter 7), but most commonly Y is regressed against X. In this case, the independent variable Y is considered to be a dependent variable in the regression analysis whereas independent variable X is considered to be an independent variable in the regression equation. In this situation, coefficients determined by the regression analysis equation are valid only for predicting values of Y using variable X as the predictor. Results of this analysis cannot be reversed and used to predict X by considering Y as the predictor without using correct inverse prediction methods (Eisenhart, 1939). This common error or misuse of regression analysis has been demonstrated by many workers; examples from earth science were given recently by Williams (1983).

If one wishes to use Y to predict X, the same data may be used if both X and Y were randomly collected (Figure 6.1, upper), but X must be treated as the dependent variable and regressed about Y, the predictor variable. Correlation coefficients will be identical in both regressions, but coefficients in the regression equation will be different in all cases except those virtually nonexistent ones when the correlation coefficient is exactly ± 1.0 and the slope (b) is exactly ± 1.0. If X was "fixed" (nonrandom) or a controlled observation (Figure 6.1, lower), regression about Y is meaningless.

Extrapolations

Often one may wish to extend predictions beyond the range of original data that were used to establish a linear relationship by regression analysis. This becomes a risky business for several reasons. Strictly speaking, it is not permitted by mathematical theory unless certain conditions are met. Nonetheless it is done fairly frequently, probably because it is better than nothing and hopefully will be better than guessing or using experience for an intuitive estimate.

All extrapolations assume that trends and relationships revealed by the regression analysis and original data base continue unchanged into the region of projected values. This may not be the case for several reasons. Linear regression models can be accurate only if we have identified adequately and have sampled the correct predictor population. Geologically, as in many natural sciences, we often have difficulty in obtaining a data set that is both representative and adequate. Commonly we are forced to use only that which is available for sampling. These data normally are obtained only in limited quantities from restricted regions; they do not represent either a random sample or an adequate, representative global sample. All too frequently, they are woefully incomplete. Therefore, our regression analysis is seriously improper and any extrapolations derived from it, by necessity, shall be extremely questionable, if not grossly erroneous.

Another common pitfall in extrapolations beyond the range of original data is that although a true linear relationship may have existed and was revealed by regression analysis within the range of original data, no assurances can be given that this will be true throughout the extrapolated zone. Many cases of this kind are well known in all areas of natural sciences. In geology, an excellent one is found in the relationship of settling velocity of a particle as a function of size (Gibbs, Matthews, and Link, 1971) and a resulting conflict between Stokes' law and Impact law.

Even in those cases where a similar linear relationship truly does continue beyond the range of original data, confidence in predictions as depicted by the confidence interval diminishes rapidly away from the center of original data (Figure 6.5). Confidence intervals are hyperbolas that are narrowest near the center of data and that recede away from the center of data. Regardless of how good an original regression was, extrapolations will always entail considerable statistical uncertainty. Of course, one can be more confident relatively if a better correlation coefficient (say > 0.90) exists than if a poor correlation coefficient (say < 0.70) exists, but it remains merely relative and even in the former case, confidence diminishes rapidly with projection beyond the original data.

A more subtle form of error encountered in extrapolations, as well as in common prediction, are those that have been determined from observations of nature (unplanned data), which are limited in extent. This is not an uncommon situation for geoscientists. Conditions in nature change constantly; some we recognize readily because we are familiar with them or expect them. Other changes remain unrecognized either because we fail to anticipate them or we are unable to analyze them properly. Yet no pre-

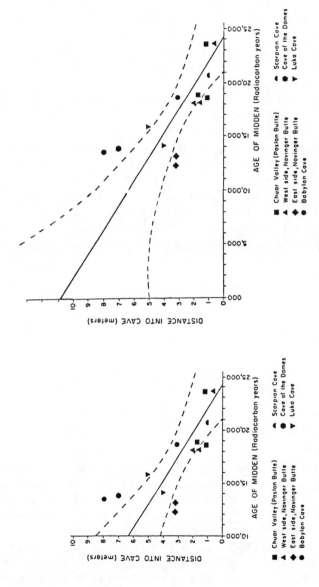

Figure 6.5 An example of prediction by linear regression beyond the range of original data. Original published data plot, regression line, and confidence interval (left, Cole and Mayer, 1982) from which the rate of erosion was concluded to be 0.45 m/10³ y and maximum expected distance for a packrat midden to be 10.8 m from a cave entrance. The original diagram has been revised here (right) to show intercept and estimated confidence interval (not calculated) near intercept.

diction or extrapolation of linear regression results is valid beyond those conditions existing for our original data base. Therefore, to provide accurate mathematical analysis, historical data, such as geologists use, must consist of a much greater amount of data than is necessary under simpler, controlled experimental conditions. Geological data must be collected over a wide range of natural conditions because these situations are uncontrolled and because they commonly involve numerous variables, both recognized and unrecognized, that may affect the data in addition to those with which we are primarily concerned. Furthermore, natural data bases must extend over a longer period of observation for three primary reasons. First, natural data are notoriously noisy data; that is, these data normally are poor quality in both accuracy and precision. Noise results from literally an infinite variety of events that are possible when numerous uncontrolled variables affect outcomes, unlike controlled experiments where we allow only one or two parameters to vary. Second, measurement of natural data commonly is poorer mechanically in uncontrolled situations than under controlled conditions. Third, data collection of unplanned or historical data commonly is not random as in designed experiments.

In summary, extrapolations fail because of inherent limitations of the data base that was used to make the projection.

Generalizations

Geoscientists typically have limited data bases with which to work, as noted previously. We often use these data, in spite of their limitations and with full knowledge that they are limited, to draw more general conclusions that go far beyond whatever limits the sampled population had. Like extrapolations, this is also an extremely risky business.

We can do this reasonably well if we are using controlled experiments and have identified fully all of the variables and have sufficient data to recognize the effects of each variable and interactions existing between variables. With uncontrolled data collection in natural situations in which we observe and gather unplanned data, however, we commonly encounter extreme difficulties in attempting to make generalizations. These difficulties are due largely to unidentified, unsuspected, and uncontrollable factors that affect our data. It also makes our geoscience data much more complex and complicated to sort out. This is one reason why progress in geological sciences has been slower than in chemical and physical sciences.

We are forced to make generalizations, nonetheless, in order to make geological progress and achieve understanding. However, it implies that we shall be wrong commonly and that we shall have to arrive at accurate generalizations predominantly through an iterative approach, rather than through individual, masterful, and brilliant deductions. We must expect old generalizations to be wrong and inaccurate more often than correct;

therefore, we constantly must review them in efforts to improve them and make them more accurate. We may also fail to make good generalizations in the geosciences because of inadequate data bases.

Causation

One must be careful not to draw unwarranted conclusions from regression analyses. A commonly implied, if not explicitly stated, conclusion seen in geological literature is derived from the mathematical necessity of considering for purposes of regression one variable (either X or Y) to be dependent and the other independent. This dependence cannot be extended from the mathematical analysis to the geoscience analysis or to actual conditions in nature. Geological dependence and independence among variables can only be determined by sound, careful geological examination of variables and factors. Mathematical analysis may and commonly does help us reach a correct geological inference by lending support to or contradicting an assumed causal interpretation.

This does not mean that we cannot use one variable to predict another variable unless a true geological dependency exists. Often in nature both variables will be related to one or more unidentified or unrecognized variables that indirectly establish those linear relationships seen in the mathematical analysis of the variables without any true geological dependency existing. These are the common "latent" variables (Box, 1966) of which we frequently know nothing. Disturbance terms and residuals in linear regression analysis reflect effects of these latent variables in addition to representing true random variations in nature. A causative linkage may exist between these latent variables and those in the regression analysis. However, regardless of latent variables and whether they are actually causal or not, regressions may still be excellent bases for predicting one variable, Y, by another, X.

We also must be cautious in drawing conclusions from regression analysis as to what relationship actually exists in nature between X and Y for another reason. Causation demands that changes in response variables be due to changes in regressor variables (Gunst and Mason, 1980) and that the predictor variable is the only one that affects the magnitude of the response variable directly. How often can we truly say this? We think more in terms of a predictor variable being a major, or even dominant factor in a causal relationship, but rarely can we ever be certain that only one, or even more variables, constitute all those which cause change in the response variable.

In addition, the situation in regression analysis is considerably different if we are using observational geological data obtained under natural conditions than if we are collecting data from a controlled experiment. In the first case, we control nothing and the data rarely are collected randomly.

Unplanned data from nature almost always have a number of latent variables that are neither suspected nor identified. Geoscientists traditionally work with natural data rather than controlled experiments and should, therefore, be concerned with these subtle aspects in mathematical analysis. Often with simple and minor modifications of data collection methods, observational data easily may be made more consistent and even meet randomization requirements. Nonetheless, when regression analysis is applied to the observation of nature and unplanned data are used, it must be with great care (Box, 1966; Snee, 1977).

Another aspect of lazy or fuzzy thinking is seen in the scientist who explores his data with regression analysis, finds a good predictor–response relationship between two variables, and then proceeds to cite this evidence as support for his conclusions as to the real-world relationship. Any data set normally will support numerous possible interpretations, hypotheses, or models of nature. Most of you are familiar with the expression "Tell me what you want to prove, and I will give you numerical support (or proof) for it," or one of the several popular paperbacks written for laymen on "How to Lie with Numbers." Their foundations lie in the fact that any data set will support many different conclusions. To draw geological conclusions properly that are valid, we must analyze thoroughly in a qualitative manner geological processes and relationships as well as variables or factors that potentially affect them *prior* to gathering data. If we have accurately analyzed the geological situation beforehand, measured the correct variables, and properly performed the mathematical analysis, the quantitative analysis clearly will support or deny our qualitative inferences. We may safely proceed in our causal interpretations of nature and the geological world only in this fashion.

Linear vs. curvilinear regressions

An assumption in linear regression modeling that commonly is ignored is that a linear relationship does exist, in fact, between two variables. Natural data commonly will not demonstrate clearly that a linear relation does or does not prevail. Neither can one assume that all natural relationships are linear because we all know phenomena that definitely are nonlinear. Thus, a question will exist normally in regression analysis as to whether the relationship demonstrated by data is most likely linear or curvilinear. Rephrased, the question is "Is the linear model correct?."

Reduction in variance. A common way to resolve this question in the past has been to test the improvement (reduction of variance, residual mean square, or sum of squares of residuals) in the regression that is due to fitting a second degree curve to the data. If improvement over the linear fit is statistically significant, presumably the original data will more likely represent a curvilinear relationship. This method has proved to be far from infallible, however. Curves of increasing degrees will always fit data better than lesser degree lines, up to degree $n - 1$ (where n is number of different data points being fitted) when the regression line will pass exactly through all data values and where variance, or residuals, will have diminished to zero. As number of data values becomes smaller, a second degree line is more likely to show a statistically significant improvement even if the true relationship is linear and thereby leads to an incorrect conclusion. Also, if the range of values predicted by the regression model is not consistently greater than the random error, prediction will commonly be of no value because the equation is fitted only to errors, even though a significant statistical test results (Draper and Smith, 1981).

Other ways of determining the correct degree of regression equation have been suggested by Kussmaul (1969) and Green (1971).

Visual inspection. A rapid visual inspection of an X vs. Y data plot (Figure 6.5, left) may suggest that the data are either linear or nonlinear. Generally, however, natural data, such as geoscience data, visually will be neither clearly for nor against linearity.

Another easy and rapid way, if it is definitive, is to examine the number of observed data points that fall beyond confidence limits calculated for the regression (Figure 6.5, left). If the number is significantly more than would be expected normally for a linear relationship and the number of data values, the relationship probably is not linear. In the case of the packrat data (Figure 6.5; Cole and Mayer, 1982), this is true; when combined with an apparent nonlinear relationship, which was suspected from visual inspection of the data plot, violation of confidence limits becomes a strong indication of a curvilinear relationship between X and Y.

Outliers. Data that are greatly dissimilar in magnitude from most observed values may arise in several ways. They may represent errors of measurement, gross operator errors, sample contamination or sampling of a mixed population, or perfectly normal variations. They also may be indicative of a nonlinear relationship. Normally residual analysis will reveal which of the many possibilities is most likely the cause of outliers within any given data set. Therefore, existence of outliers may signal nonlinearity and should always be investigated more thoroughly to determine their origin.

Hazard plots. Another easy method that often demonstrates quickly that a curvilinear relationship exists, rather than a linear one, is a hazard plot (Nelson, 1969, 1970). A hazard plot of packrat data (Figure 6.6) clearly confirms the nonlinear character of the relationship that was suspected intuitively when the original data plot was first examined visually (Figure 6.5).

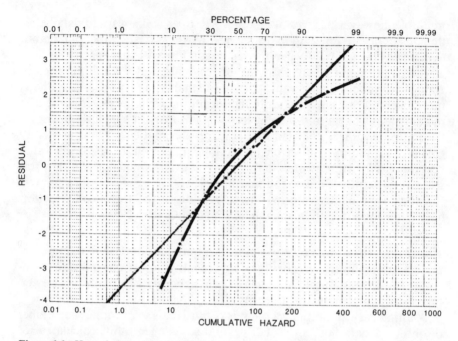

Figure 6.6 Hazard plot on normal hazard paper of packrat data (data from Cole and Mayer, 1982) clearly demonstrating the nonlinear nature of these data (dark line) rather than the assumed linear relationship (gray line).

Stratified or subset regressions of data. In another method to confirm or reject linearity of data, the data are subdivided and regression equations for subsets are determined. If subset equations are collinear or subparallel, one may reasonably conclude that a linear relationship does exist; otherwise nonlinearity probably prevails. For the packrat data, subdividing the data roughly into halves either vertically or horizontally (Figure 6.7) is inconclusive. When vertically stratified (Figure 6.7, right), the new regression equations clearly indicate nonlinearity. But horizontal stratification (Figure 6.7, left) gives equations that are not greatly different. An inspection of the upper half of the subset data, however, shows that the linear regression is extremely tenuous and could easily have been rotated 90 degrees if one data point (18,500 y, Babylon Cave) had fallen more to the left. Therefore, even the horizontally stratified subset probably should be interpreted to represent nonlinearity.

A better stratification or subset division, however, probably exists for the packrat data. The original data are an aggregate of data from seven locations. Geologically, one can argue that conditions almost certainly are not identical at these locations and, therefore, that the data should not be lumped indiscriminately. If we treat data from each of the five sites having more than one value as subsets and calculate regressions for each of them (Figure 6.8), we find a great variety in the data. Clearly, these data jointly do not demonstrate a simple linear relationship in nature. This should not be construed to mean that a linear relationship between age and distance does not exist; rather, other geological factors due to various locations of sites probably are entering the picture and should be sorted out before drawing conclusions.

A final comment on nonlinearity.   Of course, one may take an attitude that if we do not know whether or not the relationship is linear, let's be safe and fit a second degree curve to the data. This does have the advantage of reducing variance as well. However, a loss of precision results (Kussmaul, 1969); following this course of action is not desirable.

Even when all tests indicate that a linear relationship between variables is not invalidated, it does not prove that a linear relationship is correct, but merely that it is plausible and acceptable (Draper and Smith, 1981). To a certain extent, the question of adequacy of the linear model is a subjective judgment that depends on future uses of the derived regression analysis (Suich and Derringer, 1977). Can we justify the extra time and effort necessary to improve the model or to make certain that it is mathematically rigorous? As indicated earlier, the purpose of regression should govern decisions that must be made in performing, accepting, and rejecting an analysis.

Figure 6.7   Subsets of packrat data (Cole and Mayer, 1982) stratified into approximately halves vertically (right) and horizontally (left) with calculated regression equations for each.

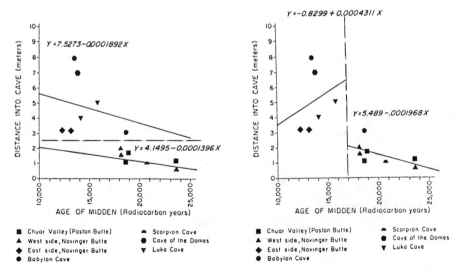

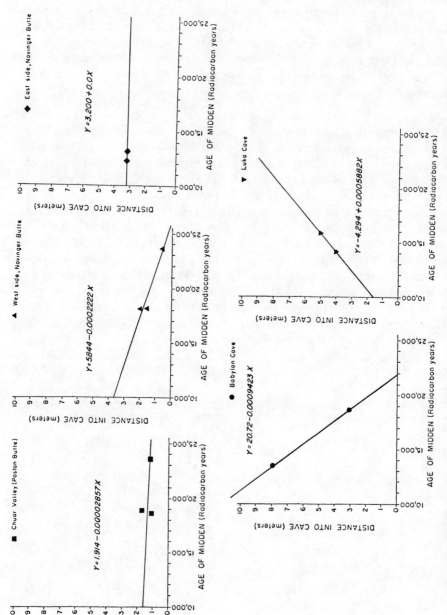

Figure 6.8 Subsets of packrat data (Cole and Mayer, 1982) stratified by locality for which two or more values were measured and regression equations calculated for each.

RESIDUAL ANALYSIS

One of the best ways currently available to differentiate linear from cur-vilinear relationships is to examine residuals. Residuals are defined as the difference between observed values and predicted values or $d_i = y_i - \hat{y}_i$ (Cox and Snell, 1968). Residuals in various forms (raw, deleted, standard-ized, and studentized) (Gunst and Mason, 1980) may be examined graph-ically in many ways (Anscombe, 1961, 1967; Anscombe and Tukey, 1963; Draper and Smith, 1966, 1981; Wooding, 1969; Chatterjee and Price, 1977; Goodall, 1983). Magnitude of residuals and their pattern are central aspects of residual analysis that can reveal adequacy of a regression anal-ysis in many different ways (Chatterjee and Price, 1977).

Here we shall look only at plots of raw residuals against X, Y, and $\hat{Y}$ (Figure 6.9). Ideally, these plots should be narrow bands of randomly placed residual values about $d = 0$ for all measured values (X, Y) or pre-

Figure 6.9 Residual plots of packrat data against predictor variable X (top), regressor variable Y (bottom left), and predicted values $\hat{Y}$ (bot-tom right). Residuals have been calculated from original data presented by Cole and Mayer (1982).

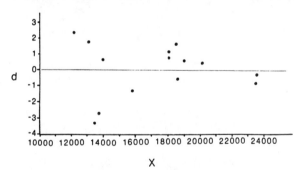

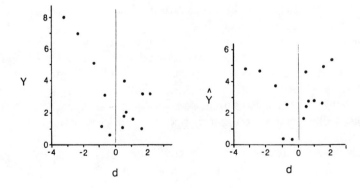

dicted values ($\hat{Y}$) (Behnken and Draper, 1972); however, deviations from this expected pattern will not always indicate failure of a regression model. For the packrat data (Figure 6.9), residuals or deviations clearly seem to be increasing as X becomes smaller (Figure 6.9, upper). When plotted against Y (Figure 6.9, middle), a linear relationship is suggested with increasingly negative residuals as Y becomes larger. However, one must be careful in using this plot (Goodall, 1983) because residuals plotted against a dependent variable can give misleading information when a correlation exists between Y and d (Jackson and Lawton, 1967). The regression analysis indicates that both residual plots here could be interpreted to suggest a nonlinear relationship. Residuals plotted against $\hat{Y}$ (Figure 6.9, bottom) clearly are increasing with larger $\hat{Y}$. Wedge patterns, shown by both X and $\hat{Y}$ plots, commonly are indicative of nonconstant variability (Goodall, 1983), that is, d is not homoscedastic.

Residual plots are easy and quick ways to examine a regression model. In addition, they often provide insight as to the nature of the data and the relationship between variables. Residual plots should be obtained routinely when linear regression analysis is performed. We should ask ourselves always, "Do these residuals seem to indicate our assumptions are wrong?" (Draper and Smith, 1981).

SUMMARY

Many problems that geoscientists encounter in regression analysis stem from the necessity of using observational or unplanned data gathered without benefit of experimental design. Historical data often will give us an apparently good linear fit to our regression model yet will be notoriously poor in predicting new data. Historical data are poorer in both accuracy and precision than data collected under controlled conditions. To provide accurate analyses mathematically, historical data must include a great quantity of data collected over a wide range of conditions, extending over a long period of observation. Failure to obtain an adequate data base frequently is the cause of poor predictive ability in an otherwise excellent linear modeling of geoscience phenomena.

Conditions in nature change, often slowly and subtly as all geologists realize. Yet it is changes that have occurred since a data set was collected and analyzed that lead to invalid present or future predictions if we fail to recognize or suspect that they are present. All potentially important variables must be identified and included, if possible, in a mathematical analysis; often they are ignored merely because data concerning them are not available.

When linear regression is used in geoscience, each of the seven assumptions inherent to the mathematical theory should be considered individually and carefully by an investigator. Although one may not be able to determine always if data do or do not meet each criterion, readers and others who subsequently use these results should be fully aware that limitations do or do not exist. Are violations of assumptions that are recognized likely to critically qualify your results? Can you find some justification, either mathematically or geologically, that will permit you to use the results in spite of these limitations?

Violations of assumed linearity of the model and assumed homoscedasticity of disturbance terms commonly are seen in the geological literature. Both generally are revealed by residual analysis, which should be made routinely by computer plots of residuals against X and $\hat{Y}$. Residual analysis is easy for nonmathematicians and requires little additional time on the part of an investigator. Often they should be included for publication along with results of regression analysis because they contain so much valuable information about the nature of the data.

Other useful techniques besides residual analysis that are helpful in evaluating assumptions and validity of regression analysis include various specific statistical tests for normality, multicollinearity, homoscedasticity or heteroscedascticity, and autocorrelation. Visual techniques include data inspection for gross errors, inspection of data plots, outliers, comparison of confidence limits with original data, time-ordered residual plots, and inspection of standard errors. Additional techniques that may be used include collecting multiple data sets, stratifying data, confluence analysis, and variance reduction testing.

When a linear model fails for a given data set, the model often may be modified easily in some manner, or the data transformed, or the data edited, or new data may be added so that violations of assumptions are no longer present. If done correctly, these changes will not invalidate the results of regression analysis.

Probably the best guidelines to follow in linear regression analysis of geological data are to use good, old common sense in each application. Use your geological knowledge and intuition to examine carefully each aspect and result of the analysis. Judge if your results are reasonable geologically. Ask yourself, "Does the mathematical solution make sense with what you might reasonably expect nature to be?"

REFERENCES

Anscombe, F. J., 1961, Examination of residuals: *Proc. 4th Berkeley Symposium on mathematical statistics and probability:* Univ. California Press, Berkeley, p. 1–36.

Anscombe, F. J., 1967, Topics in the investigation of linear relations fitted by the method of least squares: *Jour. Royal Stat. Soc., Ser. B,* v. 29, p. 1–52.

Anscombe, F. J., and Tukey, John W., 1963, The examination and analysis of residuals: *Technometrics,* v. 5, p. 141–160.

Bartlett, M. S., 1949, Fitting a straight line when both variables are subject to error: *Biometrics,* v. 5, no. 3, p. 207–212.

Behnken, Donald W., and Draper, Norman R., 1972, Residuals and their variance patterns: *Technometrics,* v. 14, p. 101–111.

Benson, M. A., 1965, Spurious correlation in hydraulics and hydrology: *Jour. Hydraul. Div. Amer. Civil Engr.,* v. 91, p. 3542–3535.

Berkson, Joseph, 1950, Are There Two Regressions?: *Jour. Amer. Stat. Assoc.,* v. 45, p. 164–180.

Bickel, P. J., 1978, Using residuals robustly. I. Tests for heteroscedasticity, nonlinearity: *Ann. Stat.,* v. 6, p. 266–291.

Blattberg, Robert C., 1973, Evaluation of the power of the Durbin–Watson statistic for non-first order serial correlation alternatives: *Rev. Econ. and Stat.,* v. 55, p. 508–515.

Box, G. E. P., 1966, Use and abuse of regression: *Technometrics,* v. 8, p. 625–629.

Box, G. E. P., and Hill, W. J., 1967, Discrimination among mechanistic models: *Technometrics,* v. 9, p. 57–71.

Champernowne, D. G., 1960, An experimental investigation of the robustness of certain procedures for estimating means and regression coefficients: *Jour. Royal Stat. Soc., London, Ser. A,* v. 123, p. 398–412.

Chatterjee, Samprit, and Price, Bertram, 1977, *Regression analysis by example:* Wiley, New York, 224 p.

Chayes, Felix, 1948, A petrographic criterion for the possible replacement origin of rocks: *Amer. Jour. Science,* v. 246, p. 413–425.

Chayes, Felix, 1949, On ratio correlation in petrography: *Jour. Geology,* v. 57, p. 239–254.

Chayes, Felix, 1960, On correlation between variables of constant sum: *Jour. Geophy. Res.,* v. 65, p. 4185–4193.

Chayes, Felix, 1962, Numerical correlation and petrographic variation: *Jour. Geology,* v. 70, p. 440–452.

Chayes, Felix, 1971, *Ratio correlation—A manual for students of petrology and geochemistry:* Univ. Chicago Press, Chicago, 99 p.

Chayes, Felix, and Kruskal, William, 1966, An approximate statistical test for correlations between proportions: *Jour. Geology,* v. 74, p. 692–702.

Chen, E. H., and Dixon, W. J., 1972, Estimates of parameters of a censored regression sample: *Jour. Amer. Stat. Assoc.,* v. 67, p. 664–671.

Cole, Kenneth L., and Mayer, Larry, 1982, Use of packrat middens to determine rates of cliff retreat in the eastern Grand Canyon, Arizona; *Geology,* v. 10, p. 597–599.

Cook, R. Dennis, and Weisberg, Sandford, 1983, Diagnostics for heteroscedasticity in regression: *Biometrika,* v. 70, p. 1–10.

Cox, David Roxbee, and Snell, E. J., 1968, A general definition of residuals: *Jour. Royal Stat. Soc., B-30,* p. 248–265.

D'Agostino, Ralph B., and Rosman, Bernard, 1974, The power of Geary's test of normality: *Biometrika,* v. 61, p. 181–184.

Davis, John C., 1973, *Statistics and data analysis in geology:* Wiley, New York, 543 p.

Draper, N. R., and Smith, H., 1966, 1981, *Applied regression analysis:* Wiley, New York, 699 p.

Durbin, J., and Watson, G. S., 1950, Testing for serial correlations in least squares regression, I.: *Biometrika,* v. 37, p. 409–428.

Durbin, J., and Watson, G. S., 1951, Testing for serial correlation in least squares regression, II.: *Biometrika,* v. 38, p. 159–178.

Durbin, J., and Watson, G. S., 1971, Testing for serial correlation in least squares regression. III.: *Biometrika,* v. 58, p. 1–19.

Dyer, Alan R., 1974, Comparisons of tests for normality with a cautionary note; *Biometrika,* v. 61, p. 185–189.

Eisenhart, C., 1939, The interpretation of certain regression methods and their use in biological and industrial research: *Ann. Math. Stat.,* v. 10, p. 162–186.

Emerson, John D., and Stoto, Michael A., 1983, Transforming data: p. 97–127, *in* Hoaglin, D. C., F. Mosteller, and J. W. Tukey, (eds.), *Understanding robust and exploratory data analysis:* Wiley, New York, 431 p.

Farrar, Donald E., and Glauber, Robert R., 1967, Multicollinearity in regression analysis: The problem revisited: *Rev. Econ. and Stat.,* v. 49, p. 92–107.

Gibbs, Ronald J., Matthews, Martin D., and Link, David A., 1971, The relationship between sphere size and settling velocity: *Jour. Sed., Pet.,* v. 41, p. 7–18.

Goodall, C., 1983, Examining residuals: p. 211–243, *in* Hoaglin, D. C., F. Mosteller, and J. W. Tukey, (eds.), *Understanding robust and exploratory data analysis:* Wiley, New York, 431 p.

Gorman, J. W., and Toman, R. J., 1966, Selection of variables for fitting equations to data: *Technometrics,* v. 8, p. 27–51.

Granger, C. W. J., and Newbold, P., 1974, Spurious regressions in econometrics: *Jour. Econometrics,* v. 2, p. 111–120.

Graybill, Franklin A., 1961, *An introduction to linear statistical models,* v. I: McGraw-Hill, New York, 372 p.

Graybill, Franklin A., 1976, *Theory and application of the linear model:* Duxbury Press, North Scituate, Mass., 698 p.

Green, J. R., 1971, Testing departure from a regression without using replication: *Technometrics,* v. 13, p. 609–615.

Gunst, Richard F., and Mason, Robert L., 1980, *Regression analysis and its application:* Marcel Dekker, New York, 397 p.

Hammerstrom, Thomas, 1981, Asymptotically optimal tests for heteroscedasticity in the general linear model: *Ann. Stat.,* v. 9, p. 368–380.

Harrison, M. J., 1975, The power of the Durbin–Watson and Geary tests: Comment and further evidence: *Rev. Econ. and Stat.,* v. 57, p. 377–379.

Harvey, A. C., and Phillips, G. D. A., 1974, A Comparison of the power of some tests for heteroskedasticity in the general linear model: *Jour. Econometrics,* v. 2, p. 307–316.

Hawkins, Douglas M., 1981, A new test for multivariate normality and homoscedasticity: *Technometrics,* v. 23, p. 105–110.

Hocking, R. R., 1976, The analysis and selection of variables in linear regression: *Technometrics,* v. 18, p. 425–438.

Hocking, R. R., 1983, Developments in linear regression methodology: 1959–1982: *Technometrics,* v. 25, p. 219–230.

Hogg, Robert V., and Randles, Ronald H., 1975, Adaptive distribution-free regression methods and their applications: *Technometrics*, v. 17, p. 399–407.

Hotelling, Harold, 1940, The selection of variates for use in prediction with some comments on the general problem of nuisance parameters: *Ann. Math. Stat.*, v. 11, p. 271–283.

Jackson, J. Edward, and Lawton, William H., 1967, Regression residual analysis: *Technometrics*, v. 9, p. 339–340.

Johnston, John, 1963, *Econometric methods:* McGraw-Hill, New York, 295 p.

Jones, Thomas A., 1979, Fitting straight lines when both variables are subject to error, I. Maximum likelihood and least-squares estimation: *Math. Geology*, v. 11, p. 1–25.

Kenney, Bernard C., 1982, Beware of spurious self-correlations: *Water Res.*, v. 18, p. 1041–1048.

Kumar, T. Krishna, 1975, Multicollinearity in regression analysis: *Rev. Econ. and Stat.*, v. 57, p. 365–366.

Kussmaul, Keith, 1969, Protection against assuming the wrong degree in polynomial regression: *Technometrics*, v. 11, p. 677–682.

Leamer, Edward E., 1973, Multicollinearity: A Bayesian interpretation: *Rev. Econ. and Stat.*, v. 55, p. 371–380.

Lindley, D. V., 1947, Regression lines and the linear functional relationship: *Supp. Jour. Royal Stat. Soc., London*, v. 9, p. 218–244.

Madansky, Albert, 1959, The fitting of straight lines when both variables are subject to error: *Jour. Amer. Stat. Assoc.*, v. 54, p. 173–205.

Mark, David M., and Church, Michael, 1977, On the misuse of regression in earth science: *Math. Geology.*, v. 9, p. 63–75.

Miller, Robert L., and Kahn, James Steven, 1962, *Statistical analysis in the Geological Sciences:* Wiley, New York, 470 p.

Nelson, Wayne, 1969, Hazard plotting for incomplete failure data: *Jour. Qual. Technology*, v. 1, p. 27–52.

Nelson, Wayne, 1970, Hazard plotting methods for analysis of life data with different failure modes: *Jour. Qual. Technology*, v. 2, p. 126–149.

O'Hagan, John, and McCabe, Brendan, 1975, Tests for the severity of multicollinearity in regression analysis: A Comment: *Review of Econ. and Stat.*, v. 57, p. 368–370.

Oja, Hannu, 1983, New tests for normality; *Biometrika*, v. 70, p. 297–299.

Olmstead, P. S., 1958, Runs determined in a sample by an arbitrary cut: *The Bell Syst. Tech. Jour.*, v. 37, p. 55–82.

Olsson, Donald M., 1979, A small-sample test for non-normality: *Jour. Qual. Technology*, v. 11, p. 95–99.

Park, Sung H., 1981, Collinearity and optimal restrictions on regression parameters for estimating responses: *Technometrics*, v. 23, p. 289–295.

Pearson, Karl, 1897, Mathematical contributions to the theory of evolution. On a form of spurious correlation which may arise when indices are used in the measurement of organs: *Proc. Royal Soc. London*, v. 60, p. 489–502.

Pericchi, L. R., 1981, A Bayesian approach to transformations to normality: *Biometrika*, v. 68, p. 35–43.

Pesaran, M. H., and Slater, L. J., 1980, *Dynamic regression: Theory and algorithms:* Halsted Press (Wiley), New York, 355 p.

Pierce, Donald A., and Gray, Robert J., 1982, Testing normality of errors in regression models: *Biometrika,* v. 69, p. 233–236.

Plackett, R. L., 1972, The discovery of the method of least squares: *Biometrika,* v. 59, p. 239–251.

Poole, Michael A., and O'Farrell, Patrick N., 1971, The assumptions of the linear regression model: *Trans. Inst. British Geog.,* no. 52, p. 145–158.

Prestegaard, Karen L., 1983, Variables influencing water-surface slopes in gravel-bed streams at bankful stage: *Geol. Soc. Amer. Bull.,* v. 94, p. 673–678.

Scheffé, Henry, 1959, *The analysis of variance:* Wiley, New York, 465 p.

Schmidt, Peter, and Guilkey, David K., 1975, Some further evidence on the power of the Durbin–Watson and Geary tests: *Rev. Econ. and Stat.,* v. 57, p. 379–382.

Seal, Hilary L., 1967, The historical development of the Gauss linear model: *Biometrika,* v. 54, p. 1–24.

Seber, G. A. F., 1977, *Linear regression analysis:* Wiley, New York, 457 p.

Shapiro, S. S., and Wilk, M. B., 1965, An analysis of variance test for normality (complete samples): *Biometrika,* v. 52, p. 591–611.

Shapiro, S. S., Wilk, M. B., and Chen, H. J., 1968, A comparative study of various tests for normality: *Jour. Amer. Stat. Assoc.,* v. 63, p. 1343–1372.

Snedecor, George W., 1956, *Statistical methods:* Iowa State College Press, Ames, 507 p.

Snee, Ronald D., 1977, Validation of regression models: Methods and examples: *Technometrics* v. 19, p. 415–428.

Stigler, Stephen M., 1981, Gauss and the invention of least squares: *Ann. Stat.,* v. 9, p. 465–474.

Stransky, Terry, 1984, Comment on "Improper use of regression equations in earth sciences": *Geology,* v. 12, p. 125–126.

Suich, Ronald, and Derringer, George C., 1977, Is the regression equation adequate? One criterion: *Technometrics,* v. 19, p. 213–216.

Swindel, Benee F., and Bower, David R., 1972, Rounding errors in the independent variables in a general linear model: *Technometrics,* v. 14, p. 215–218.

Tarter, Michael E., and Kowalski, Charles J., 1972, A new test for and class of transformations to normality: *Technometrics,* v. 14, p. 735–744.

Tsutakawa, R. K., and Hewett, J. E., 1978, Comparison of two regression lines over a finite length: *Biometrics,* v. 34, p. 391–398.

Tukey, John W., 1960, Conclusions vs. decisions: *Technometrics,* v. 2, p. 423–433.

Vistelius, Andrew B., and Sarmanov, Oleg V., 1961, On the correlation between percentage values: Major component correlation in ferromagnesium micas: *Jour. Geology,* v. 69, p. 145–153.

Wald, Abraham, 1940, The fitting of straight lines if both variables are subject to error: *Ann. Math. Stat.,* v. 11, p. 284–300.

Wichers, C. Robert, 1975, The detection of multicollinearity: A comment: *Rev. Econ. and Stat.,* v. 57, p. 366–368.

Winsor, Charles P., 1946, Which Regression?: *Biometrics,* v. 2, p. 101–109.

Williams, Garnett P., 1983, Improper use of regression equations in earth sciences: *Geology,* v. 11, p. 195–197.

Willan, Andrew R., and Watts, Donald G., 1978, Meaningful multicollinearity measures: *Technometrics,* v. 20, p. 407–412.

Wolfe, Douglas A., 1977, A distribution-free test for related correlation coefficients: *Technometrics,* v. 19, p. 507–509.

Wonnacott, Thomas H., and Wonnacott, Ronald J., 1981, *Regression: A second course in statistics:* Wiley, New York, 548 p.

Wooding, W. M., 1969, The Computation and use of residuals in the analysis of experimental data: *Jour. Qual. Technology,* v. 1, p. 175–188.

Yale, Corallee, and Forsythe, Alan B., 1976, Winsorized regression: *Technometrics,* v. 18, p. 291–300.

Yalin, M. S., and Kamphuis, J. W., 1971, Theory of dimensions and spurious correlation: *Jour. Hydraul. Res., I. A. H.,* v. 9, p. 249–265.

7

Fitting Straight Lines in the Earth Sciences

Brent M. Troutman
and
Garnett P. Williams

The straight line probably is the most important and widely used type of statistical relation in earth sciences. It is the simplest and most understandable way to describe the relation between two variables. Many people believe there are only two ways to fit a straight line to data: The "eye" or freehand method, if a general approximation is adequate, and least squares, if one wants to be "scientific" and "precise." In fact, there are many ways of fitting a straight line. The purposes of this chapter are to discuss some of the various methods, their advantages and disadvantages, and their appropriate applications.

Some purposes in fitting a straight line might be to:

1. Summarize data, or represent a system of points, by a single equation.
2. Try to find an underlying physical law or unique functional relation between two variables, confirm or refute a theoretical relation, or help choose a theoretical or mathematical model.
3. Provide a basis for predicting values of one variable, given values of another.
4. Objectively compare several groups of data in terms of constants in best-fit equations for each group.
5. Calibrate a new instrument in terms of an established one.

These purposes are not mutually exclusive; however, a given line-fitting technique can be more appropriate for some purposes than for others.

Before fitting a straight line to any data set, the data *must* be plotted. This ensures that the pattern of plotted points is visually linear. (In the modern computer era, some investigators are tempted simply to put all data on the computer, fit a computer-determined straight line to the data, and draw inferences from the resulting statistics without ever having confirmed that the variables show a definable relation and that a straight line is appropriate!) Also, subsequent use of the fitted equation should be restricted to the range of values of the variables used in the derivation of the equation (see Mann, Chapter 6).

In this chapter, we consider only one independent variable. The first part of the chapter describes some line-fitting techniques that have been popular in the scientific literature, although some have not received much attention in earth sciences. We then illustrate several of these techniques by an application to the same set of data. Next, we introduce statistical analysis as a means of determining whether certain techniques may be more appropriate than others under specified conditions. Finally, we make recommendations on when to apply the various line-fitting procedures.

THE EQUATION OF A STRAIGHT LINE

The equation of the straight line relating variables X and Y is

$$Y = a + bX \tag{1}$$

The constant a is the value of Y given by the straight line at $X = 0$. The constant b is the slope of the line and shows the number of units increase (or decrease) in Y that accompanies an increase in one unit of X. Straight-line fitting is used simply to determine the values of the constants (parameter estimates) a and b.

As it stands, (1) is a simple algebraic relation that allows for no deviation of points from the line represented by the equation. Points in real-world data sets almost always exhibit scatter, or deviations from the line, and it is this feature that introduces uncertainty into straight-line fitting procedures.

In the next section, we present some methods that may be used to fit a straight line to a set of (X, Y) pairs with scatter. At this point these methods are treated simply as mechanical exercises, with no guidelines given as to how one should select an appropriate technique. In subsequent sections, we discuss statistical considerations and how these considerations can assist in determining which techniques are more appropriate. In statistical analysis, (1) is modified to include "error" terms that explicitly represent the deviations from the straight-line model. The nature of these errors often dictates which procedures will be better.

BASIC METHODS OF FITTING A STRAIGHT LINE

Eye

Drawing a straight line subjectively by eye, using a transparent straight-edge, has been a popular technique, mainly because it is quick and easy. Once the line has been drawn, the constants a and b (and hence the equation of the line) can be determined graphically or from any two points (preferably as far apart as possible) on the line. (When the two points arbitrarily are chosen in advance and the line deliberately drawn through these points, the method is called the method of selected points.)

If the scales of the graph are logarithmic rather than arithmetic, the logs of X and Y must be used in computing the constants. The intercept a when read from the graph on log paper is the value of Y given by the straight line at $X = 1$.

Averages

This method, sometimes called "semiaverages" or "group averages by residual summation," consists of (1) dividing the data into two equal or nearly equal groups, usually according to ascending values of X, (2) calculating the average X (symbol $\bar{X}$) and average Y ($\bar{Y}$) within each group, (3) plotting these two mean points on the scatter diagram, and (4) connecting them with a straight line. The constants a and b can be determined as above.

A variant of this technique is to (1) arrange the entire data set according to ascending values of X, (2) divide the data into several groups (rather than just two groups), (3) compute and plot the ($\bar{X}$, $\bar{Y}$) point for each group, and (4) fit a straight line by eye to the several averaged points. Use of average values usually results in vastly reduced scatter of plotted (averaged) points, but the true scatter is still reflected by the original data.

Least squares (OLS)

Ordinary least squares (OLS) produces the line having the minimum sum of squared deviations between the data points and the line. This method frequently is used for prediction purposes, as discussed below. If Y is the variable to be predicted, the deviations are taken in the vertical direction, shown as line AB in Figure 7.1. If X were the variable to be predicted and we wished to regress X on Y, the appropriate deviation would be the hor-

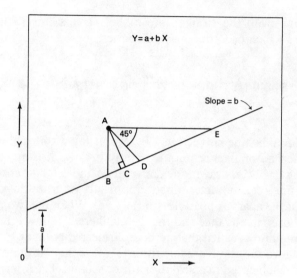

Figure 7.1 Straight line with various deviations or distances between a data point (*A*) and the line.

izontal distance *AE*. Minimizing the sum of squared horizontal deviations usually gives a different best-fit line from that obtained by minimizing the sum of squared vertical deviations. OLS is the technique normally taught in introductory statistics courses.

Certain basic statistics for calculating *a* and *b* are used not only with the least-squares method but also with several other methods (Hirsch and Gilroy, 1984). These statistics are:

$$S_x = \left(\frac{1}{N} \Sigma (X_i - \bar{X})^2 \right)^{1/2} \tag{2}$$

$$S_y = \left(\frac{1}{N} \Sigma (Y_i - \bar{Y})^2 \right)^{1/2} \tag{3}$$

$$S_{xy} = \frac{1}{N} \Sigma (X_i - \bar{X})(Y_i - \bar{Y}) \tag{4}$$

and

$$r = \frac{S_{xy}}{S_x S_y} \tag{5}$$

where *N* is the sample size, Σ means summation, $\bar{X}$ is the mean and S_x is the standard deviation of the sample of *X* values, $\bar{Y}$ and S_y are the mean and standard deviation of the *Y* values, S_{xy} is the covariance, the subscripts

i refer to individual observations of the associated variable, and r is the well-known correlation coefficient. The correlation coefficient is just a scaled version of the covariance, and both simply measure the association between X and Y.

Of the line-fitting methods that use the above statistics, the way in which the slope b is calculated varies from one line-fitting method to another. Once b is determined, the intercept a is easily computed. The lines that use the statistics of (2) to (5) all pass through the centroid of the data $(\bar{X}, \bar{Y})$. Therefore, regardless of how the line is fitted, the computed values of $\bar{X}, \bar{Y}$ and b can be inserted into (1) to solve for a. This gives

$$a = \bar{Y} - b\bar{X} \qquad (6)$$

when Y is the dependent variable.

For the least-squares method with Y as the dependent variable, the value of b is simply

$$b = r(S_y/S_x) \qquad (7)$$

Structural line

The structural line can occupy any position between the two OLS lines, depending on the value of a parameter λ. λ reflects the relative errors in X and Y and will be precisely defined later in the chapter.

The slope of a structural line is defined by

$$b = \frac{S_y^2 - \lambda S_x^2 + ([S_y^2 - \lambda S_x^2]^2 + 4\lambda S_{xy}^2)^{1/2}}{2S_{xy}} \qquad (8)$$

where λ is a nonnegative number (see, for example, Madansky, 1959). The intercept of the line is again given by (6).

Equation 8 often reduces to much simpler and less formidable forms, depending on the value of λ. In addition, when λ tends to infinity, (8) yields the OLS line that minimizes the sum of squared vertical deviations. When $\lambda = 0$, the slope (equation 8) is identical to the OLS line computed by minimizing the sum of squared horizontal deviations. [That is, in an OLS fit of X on Y the computed slope (b' in $X = a' + b'Y$) is $b' = S_{xy}/S_y^2$. If this equation for the line is inverted to the form $Y = a + bX$, we get $b = 1/b' = S_y^2/S_{xy}$, which is (8) with $\lambda = 0$.] Two special cases of structural analysis are the least normal squares line and the reduced major axis, discussed next.

Least normal squares (LNS)

Minimizing the sum of the squared shortest (perpendicular) distances between the line and the data points (line AC in Figure 7.1) produces the least normal squares line (sometimes called the major axis, principal axis, or Pearson's, 1901, line of closest fit). For $r \neq 0$, the slope b of this line is

$$b = \frac{-A + (r^2 + A^2)^{1/2}}{r} \tag{9}$$

where

$$A = (\tfrac{1}{2}) \left(\frac{S_x}{S_y} - \frac{S_y}{S_x} \right) \tag{10}$$

Equation 6 then provides the intercept, a. The slope b of (9) is the special case of (8) when $\lambda = 1$.

Reduced major axis (RMA)

This method or line, also known as the unique line of organic correlation, minimizes the sum of areas of right triangles formed between the data points and the best-fit line, with the legs of the triangles drawn parallel to the X and Y axes of the graph (BAE in Figure 7.1). (In this geometrical sense, the OLS and LNS methods can be viewed as minimizing the sum of areas of squares, one side of which is the point's deviation.) The triangular area for any data point and RMA-fitted line is $\tfrac{1}{2}$ (base $\times$ height) = $\tfrac{1}{2}$ (vertical deviation $\times$ horizontal deviation). In Figure 7.1 this would be $\tfrac{1}{2}$ of ($AB \times AE$).

The slope b of the RMA line is simply

$$b = \pm(S_y/S_x) \tag{11}$$

in which b is given the sign of the correlation coefficient. The slope in (11) is the special case of (8) when λ is numerically equal to S_y^2/S_x^2.

Other methods

Various other methods of fitting straight lines to data have been proposed over the years but have not become popular. Examples are (a) the line bisecting the OLS fits of Y on X and X on Y; (b) the minimization of the sum of squares of each 45° line between a data point's horizontal and vertical deviations to the best-fit line (line AD of Figure 7.1) (see, for instance, Zucker, 1947); and (c) the line passing through the center of mass of the data ($\overline{X}$, $\overline{Y}$) and one other point known to be on the line, as in some cases the origin ($X = 0$ and $Y = 0$).

APPLICATION OF SELECTED METHODS TO THE SAME DATA SET

Figure 7.2 shows most of these lines fitted to a hypothetical data set that represents the regional variation of quartz in a sedimentary rock stratum.

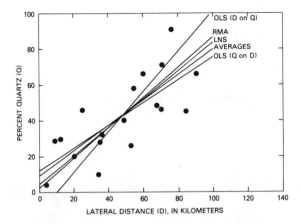

Figure 7.2 Hypothetical data for a sedimentary rock stratum showing change in percent quartz with lateral distance, and fitted lines determined by various methods.

Let D = horizontal distance from an arbitrary base point, in kilometers, and Q = percent quartz in the mineralogical composition. In Figure 7.2 there is a general linear trend on arithmetic scales.

Suppose that both D and Q are uncontrolled and measured with error. From a causality viewpoint, Figure 7.2 is the most logical arrangement, with percent quartz plotted as the dependent variable on the vertical scale. However, percent quartz probably depends on other variables and only has an association with distance. Also, under certain circumstances, one might wish to predict distance from a knowledge of percent quartz. Thus either variable conceivably could be "dependent," according to the circumstances.

The means of the data are 47.28 km (D) and 42.00% (Q); standard deviations are 25.86 km (D) and 21.79% (Q); and $r = 0.746$. The calculated equations in Figure 7.2 are:

Averages	$Q = 9.40 + 0.69D$	(12)
OLS, Q on D:	$Q = 12.12 + 0.63D$	(13)
OLS, D on Q:	$D = 9.90 + 0.89Q$	(14)
LNS:	$Q = 4.60 + 0.79D$	(15)
RMA:	$Q = 2.16 + 0.84D$	(16)

Only when all data points plot exactly on a line, that is, when $r = 1$ (or -1), will all five lines be identical. If $S_y = S_x$, LNS and RMA give the same line, and the slope of this line will be 1 (or -1) (see equations 9 and 11). Otherwise, all five lines will be different.

The two OLS lines represent the extremes of possible fits, and their divergence increases as the correlation between variables becomes nearer to zero. Also, disagreement between lines increases as distance from the centroid $(\overline{X}, \overline{Y})$ increases. As correlation improves, $r \to 1$ (or $r \to -1$), and the OLS lines converge on the RMA line (see equations 7 and 11). Thus the RMA line may be viewed as the limiting line, as $r \to 1$ (or $r \to -1$), of the two OLS lines.

Most of the other line-fitting methods discussed in this chapter also give different lines. In fitting straight lines to data, we therefore have the dilemma of which method to use.

WEIGHTING METHODS AND TRANSFORMATIONS OF DATA

There are several other procedures that may be described as generalizations or refinements of the methods discussed above. These procedures are ways of weighting, or assigning importance to, the data points. They could be applied within the context of any of the methods discussed previously.

Transformation of data

Many sets of data plot as curves on arithmetic scales but can be rectified to a straight-line pattern if transformed in some way. The most common transformation is to take logarithms of X and (or) Y. Other common transformations are $1/X$, $1/Y$, X^2, Y^2, X/Y, Y/X, log (Y minus a constant), and many others. All straight-line fitting procedures we have discussed may be applied to transformed data.

In addition to possibly producing a straight-line plot, a data transformation gives different weights to X and Y values with different magnitudes. For example, as shown below, minimizing the sum of squared deviations for log Y vs. log X gives less weight to large values of X and Y than does minimizing the sum of squared deviations for Y vs. X. This commonly is very appropriate, because in geologic data large values tend to be subject to larger absolute errors of measurement than small values; that is, percentage (relative) errors tend to be more nearly equal than absolute errors.

A simple example shows the effect of logarithmic and reciprocal transformations. Suppose three replicate measurements of Y at a given X yield values of 10, 100, and 1000. An estimate of the expected value of Y would be the arithmetic average, or 370. If, however, we transform to logs, the average log (base 10) for the three observations would be $(1 + 2 + 3)/3 = 2.00$. The antilog of this average log is 100, only about one-fourth the arithmetic average of 370. The transformation has given less weight to the larger arithmetic values (such as 1000 in this example). Similarly, if our transformation of the same data as above consisted of taking $1/Y$, the individual values of 1/10, 1/100 and 1/1000 would yield an average $1/Y$ of 0.037. Taking $1/Y = 0.037$ and solving for Y gives $Y = 27$. Here, the transformation has produced a value of Y that is less than one-tenth the arithmetic average of 370. Again, the transformation has given less weight to the larger values of the original data. On a graph of untransformed (original) data for these types of cases, the line that best fits the transformed data (after recalculation into original coordinates) plots lower than that for the untransformed data. In general, predicting logs or reciprocals is quite different than predicting the untransformed values of a variable. Jansson (1985) discusses several key problems in log transformations.

Direct weighting and weighted least squares (WLS)

As shown in the preceding paragraphs, transformations have the effect of weighting the data. Instead of applying a transformation, different weights can be assigned directly to the data points before performing the line fitting. These weights must be specified by the researcher. Terms with larger weights have more influence in the fitting procedure. The usual justification for giving certain data more weight than others is that some observations may be considered less reliable, or subject to more error, than others; weighting allows the researcher to give such observations less influence in fitting a straight line. This variation in reliability may occur, for example, if data are collected by different experimenters or if different instruments are used in collecting the data.

One well-known procedure that involves weighting in fitting a straight line is weighted least squares (WLS), a generalization of OLS. In OLS, a sum of squared deviations from the line is minimized; WLS is similar, but each squared deviation is multiplied by a weight (w_i) before summing (Draper and Smith, 1981, Section 2.11). Equations 6 and 7 for computing the OLS slope and intercept also are used with WLS; the only difference is that $(X_i - \bar{X})$ and $(Y_i - \bar{Y})$ in (2), (3), and (4) are multiplied by $w_i^{1/2}$, N is replaced by Σw_i, and $\bar{X}$ (or $\bar{Y}$) is computed by $\Sigma w_i X_i / \Sigma w_i$ (or $\Sigma w_i Y_i / \Sigma w_i$). Statistical packages for WLS are available from many sources; an example is Freund and Littell (1981).

WLS usually is applied only in the context of statistical analysis because such analysis gives guidelines on how the weights should be chosen. The weights generally are chosen by seeking a pattern in the residuals, where the residuals are measured from an OLS line. Where repeat measurements of Y are not made, the usual approach is to seek a model for the variance of the residuals. For example, the variance of the residuals might increase or decrease with (i.e., be a function of) X. In this case, a plot of residuals versus X would tend to show increasing or decreasing scatter of the residuals with X. Draper and Smith (1981, p. 111 and 237) and Chatterjee and Price (1977, p. 101–122) discuss this general problem.

Robust methods

Robust methods do not allow any single data point (such as an outlier) or group of points to have a disproportionate influence on the computed values of slope and intercept of a fitted line. Thus, robust methods in a sense are another way of weighting the data points. In OLS, where points are not weighted, a point with a large vertical deviation can contribute a large proportion to the sum of the squared deviations. This causes a minimization procedure to try to move the line toward this point. Thus the usual OLS estimates of slope and intercept are particularly susceptible to the potentially disastrous effects of one or two points that do not conform to the pattern of the other points (Huber, 1981, section 7.1).

One general type of robust estimate (the "M estimate") that has been proposed to resolve this problem is obtained by minimizing the sum of a selected function of the vertical distance (residual) between each point and the line. (Choosing the selected function to be the square of the residual gives the OLS estimate.) A popular technique uses the absolute value of each residual, yielding the least absolute deviation (LAD) estimate. Thus, LAD is similar to OLS, except that LAD minimizes the sum of absolute values of the residuals instead of the sum of squared residuals.

Huber (1973) proposed using a function that combines these two estimators; his estimator essentially squares the residual if it is small and takes the absolute value if it is large. Other types of robust estimates (including the so-called "L estimates" and "R estimates") have been proposed. Huynh (1982), Huber (1981), and Hogg (1979) give further discussion.

The slope of the fitted line with robust methods must be determined iteratively. An initial slope is determined by OLS or by some other method; weights based on this fitted line then are computed; the WLS technique then is used with these weights to calculate a new slope; the entire process is repeated until the computed slope converges to a final value.

STATISTICAL FRAMEWORK FOR FITTING STRAIGHT LINES

Why a statistical framework?

Thus far we have presented several techniques for fitting straight lines. Each technique may be applied to any set of data consisting of paired observations on two variables. The techniques are simply algebraic or mechanical, and no structure has been imposed that places the procedures within the realm of statistical analysis. The reason we may wish to place a problem into a statistical framework is that doing so provides an objective means of answering the following questions: Which technique is best? After applying a particular technique, how good is the line that is computed? Statistical analysis gives definite and objective meanings to the words "best" and "good."

What makes an analysis statistical?

To make an analysis statistical, we first postulate some underlying framework or set of assumptions that characterizes the properties of the data set. This framework almost always includes an assumption that some or all of the variables of interest are *random* variables. A random variable is one for which a precise value can't be predicted but which instead is characterized by a probability distribution. We then define the straight line of interest in terms of these probability distributions. The line we wish to estimate is called a *population* line or "true" line; it is fixed but unknown. The intercept and slope of the population line are known as parameters. (Practitioners sometimes erroneously call the variables X and Y parameters.) The N pairs of observations constitute a random sample that is used to estimate the population line (i.e., the parameters). Another set of measured pairs would lead to a different sample line; in other words, the sample line is itself random. The "goodness" of a procedure for fitting the straight line is then judged by how close the sample line is, on the average, to the true population line. Closeness is measured, for example, by the standard error of the slope (b) and intercept (a); we will not go into computational details of such measures in this chapter.

Definitions of regression, structural analysis, and functional analysis

To fit a straight line within a statistical framework, it is necessary first to specify exactly what population line is of interest. Various population lines may describe a relation between two random variables, X and Y; the line to estimate depends primarily on how the line will be used after it is computed. Three common and useful population lines are the regression relation, the structural relation, and the functional relation.

The regression of Y on X is defined to be the line that gives the relation between X and the mean (or expected) value of Y, given X (Kendall and Stuart, 1967, chapter 28; Graybill, 1976, chapter 5). That is, for any X, Y is assumed to have a probability distribution, and the mean value of this distribution defines the regression line. In this situation, Y is called the dependent and X the independent variable. The regression relationship between X and Y commonly is written as

$$Y = \beta_0 + \beta_1 X + e \qquad (17)$$

where $\beta_0 + \beta_1 X$ is the population regression line and e represents the "error," or deviation, of Y from the straight line.

An analogous definition holds for the regression of X on Y, for which X is the dependent variable. This regression line is different from that of Y on X (Figure 7.2). To define either line, the dependent variable must be random. The independent variable, however, may be either random or nonrandom.

In structural and functional analysis, it is assumed that there are two other variables (X^* and Y^*) that are error-free counterparts of X and Y; the ultimate goal is to make inferences about the relationship between X^* and Y^*, *not* about the relationship between X and Y. Because it is virtually impossible to make measurements without error, X^* and Y^* are not observable. All we have are the erroneous measurements X and Y and, of course, we hope that X is close to X^* and Y is close to Y^*. The errors in trying to observe X^* and Y^* may be any combination of (a) experimental or measurement error, (b) random variability, which is unpredictable, uncontrollable, and unmeasurable, and (c) the effect of other variables not included in the model.

Usually, it is assumed that X^* and Y^* are perfectly linearly related:

$$Y^* = \alpha_0 + \alpha_1 X^* \tag{18}$$

where α_0 is the intercept and α_1 the slope of the line. Notice that the assumption of a "perfect" relationship means that there is no error term as there is in the relationship between X and Y in (17). The errors in attempting to measure X^* and Y^* will be denoted by u and v, respectively:

$$X = X^* + u \tag{19}$$

and

$$Y = Y^* + v \tag{20}$$

When X^* and Y^* may themselves be random, (18) is usually called a structural relation between X^* and Y^*; when X^* and Y^* are nonrandom, it is called a functional relation. Examples of the two situations in the earth sciences may be found in the work by Jones (1979, p. 5–6). See also Kendall and Stuart (1967, p. 375–378) and Madansky (1959, p. 175–176). Because in this chapter we are primarily interested in the situation for which X^* and Y^* are random, subsequent discussion is limited to structural analysis.

To summarize, in regression we estimate the mean value of Y given X, or the mean value of X given Y. In structural analysis, we are interested in the linear relationship between the (unobservable) error-free variables X^* and Y^*.

Confusion often arises concerning the relationship of regression and structural analysis and the techniques for line fitting presented previously. Regression and structural analysis are not really single "techniques" at all; rather, each is *any* procedure (including, conceivably, all the techniques discussed in this chapter) that is applied with the specific intention of estimating β_0 and β_1 (for regression) or α_0 and α_1 (for structural analysis). Although we may apply any method discussed so far to estimate the slope and intercept, each technique must be evaluated in terms of how close, on the average, the estimated line is to the population line. One such measure of "closeness" would be the mean-squared error of the slope and intercept estimates.

Estimation of parameters

Optimal procedures have been derived for estimating the regression and structural lines. Many of these procedures are based on the principle of maximum likelihood (ML) (Mood, Graybill, and Boes, 1974, section VII, 2.2). In regression analysis, if the errors e in equation 17 are normally distributed with mean zero and constant variance (variance that is the same for all X), and if the errors are independent of the X's and of each other, then ML reduces to ordinary least squares; if the errors are double-exponentially distributed rather than normally distributed, ML reduces to least absolute deviation; if the errors are normally distributed and if the weights w_i are inversely proportional to the variance of Y_i, ML reduces to weighted least squares.

In structural analysis, a value for λ is required:

$$\lambda = \text{var}(v)/\text{var}(u) \tag{21}$$

where $\text{var}(v)$ is the variance (or squared standard deviation) of the probability distribution of v, and $\text{var}(u)$ is similarly defined. If the errors u and v in (19) and (20) are normally distributed with mean zero and constant variance, if the errors are independent of the X^*'s and Y^*'s and of each other, and if λ is known, then the ML estimator is the structural line (Madansky, 1959; Kendall and Stuart, 1967; Osterkamp, McNellis and Jordan, 1978), the slope of which is given in (8).

The main problem in using (8) to estimate the slope in structural analysis is that λ is assumed to be known. In most practical applications, a value for λ must be assumed or estimated (see Mark and Church, 1977). For example, when RMA is used, the investigator assumes that S_y^2/S_x^2 is a good approximation of $\text{var}(v)/\text{var}(u)$, so that $\lambda = S_y^2/S_x^2$ and (8) reduces to (11). For LNS, the assumption is that the variance of errors u and v is the same for both variables, that is, $\text{var}(v) = \text{var}(u)$ and $\lambda = 1$. Lakshminarayanan and Gunst (1984) discuss sensitivity of the slope estimator (8) to erroneous selection of λ.

Estimators for the structural line also may be obtained under assumptions other than those given above. For example, rather than assuming λ is known, it may be assumed that either $\text{var}(u)$ or $\text{var}(v)$ is known, or that both of these are known. (The latter assumption is stronger than assuming that only λ is known.) Also, estimators are available for the situation where one has replicate observations on X (for a given X^*) and Y (for a given Y^*). Finally, another procedure for estimating the structural line is a variation of the method of averages discussed in this chapter. Madansky (1959) gives computational details for these situations.

Prediction and estimation: What is the purpose of fitting the line?

The regression of Y on X, the regression of X on Y, and the structural line are three distinct population lines that describe the relationship between two random variables (e.g., Figure 7.2). Before any procedure for fitting a straight line within a statistical framework is applied, the investigator must decide which line is really of interest. The appropriate line is not always obvious, but here is a rule of thumb: To *predict* the value of Y (or X) associated with a given observed value of X (or Y), that is, to estimate the dependent variable's value that statistically is likely to have smallest error, the regression of Y on X (or X on Y) is in general the best line to use. This regression line usually is estimated by OLS or a related technique, such as WLS. If, on the other hand, one is more interested in (a) *estimating* the true underlying linear relationship or association between two variables (both of which may be observable only with error), especially when there is no clear causality, or (b) estimating the intercept and slope that are more likely to be physically meaningful or realistic, or (c) comparing the fitted line to a theoretical line or to various other fitted lines, then the structural line (RMA, LNS, or other) is probably the best line to use (see Madansky, 1959; Mark and Church, 1977). Thus, the structural line is more likely to be appropriate when one is interested in the line relating the true, or error-free, variables X^* and Y^*, and the regression line is more appropriate if one intends to use observed values of the independent variable (which are necessarily subject to error) in the fitted equation to obtain predictions of the dependent variable.

This rule of thumb means that investigators should not routinely apply least squares every time they want to fit a straight line to data. Instead, they must first choose a primary purpose for the line. The purposes just mentioned are all desirable, of course, and everyone naturally would prefer a single line or method that fulfills all of them. Unfortunately, however, such a single, supreme method or line does not exist (Zucker, 1948, p. 49; Greenall, 1949, p. 40). OLS in many cases is used inappropriately, namely, in situations where an estimate of the structural line should be used (Till, 1973; Mark and Church, 1977). See also Troutman (1982, 1983) for a discussion of problems associated with using OLS to estimate parameters in precipitation-runoff regressions when the independent variable (precipitation) is subject to error.

Regression is appropriate even if the independent variable is subject to error, provided that the values of the independent variable to be used in the computed line are subject to the same type of error as the values used to obtain the estimates of slope and intercept; this condition generally is fulfilled in a prediction situation.

Interpretation of lines computed with transformed data

Care must be exercised in interpreting lines that are obtained using transformed data. Consider, for example, the OLS line computed with logarithmically transformed data, log Y_i and log X_i. Suppose that for a given X we are interested in estimating the mean value of Y itself, not just the mean of log Y. Taking the antilog of the value given by the fitted line does not work (see the section, "Transformation of Data"). If we can assume that log Y has a normal distribution (for each X), then one way to obtain mean (Y) is to apply a simple correction to the value obtained from the line before taking the antilog. The formula in natural logarithms is mean (Y) = antilog (mean [ln Y] + ½ variance [ln Y]), where mean [ln Y] is obtained from the fitted line at the given X and variance [ln Y] is the regression variance of points around the line (see Aitchison and Brown, 1957; Mejía and Rodriguez-Iturbe, 1974; Miller, 1984; Jansson, 1985). Taking the antilog of the value from the fitted line without applying the correction really estimates the *median* (50th percentile) of Y. If log Y has a normal distribution, Y itself has a log-normal distribution, and the median is less than the mean for this distribution.

This leads to another consideration in line fitting: Exactly what properties of the probability distribution of Y really are of interest? The discussion in the previous paragraph deals with two different measures of central tendency: mean (arithmetic average) and median. The latter often is more appropriate for skewed data, although the user must decide in particular situations which measure is better. Other measures of central tendency include the mode (value of Y corresponding to the peak of the frequency distribution), the geometric mean (antilog of the mean log Y), and the harmonic mean (the reciprocal of the mean $1/Y$). The geometric mean corresponds to the median when log Y is normally distributed, and the harmonic mean of Y is estimated by the OLS line obtained by fitting $1/Y$ to X. There are no simple rules to be followed when using transformations; be cautious when interpreting lines fitted to transformed data.

Analysis after the fit, and model building

By placing a line-fitting problem into a statistical framework, various objective measures of goodness of the line become available. These measures include the coefficient of determination (r^2), the standard deviation of the points around the line, and the standard errors of the slope and intercept (Imbrie, 1956; Madansky, 1959; Draper and Smith, 1981). An additional essential step is verifying that any assumptions made in the analysis do indeed hold. For example, in applying OLS it is assumed that the errors have constant variance around the line and, ideally, that the errors are normally distributed. The best way to verify any assumptions is to examine the residuals (deviations of the points from the fitted line). Visual examination usually is sufficient. For specific techniques of residual examination see Draper and Smith (1981, Chapter 3) and Cook and Weisberg (1982).

One or more of the assumptions may not be appropriate. One then has a choice either of ignoring the fact that the assumptions are violated and proceeding with the given line-fitting technique anyway or of applying a different technique that more nearly reflects the properties of the data. Violation of certain assumptions may not be of much consequence; for instance, an OLS line has desirable properties even if the errors are not normally distributed. In general, though, it is worthwhile to amend the technique when assumptions are violated. For example, if a researcher performs an OLS fit and then finds that the residuals do not appear to have constant variance (for example, if the residuals tend to exhibit more variability as X increases), he could then go back and attempt to incorporate the unequal variances by refitting the line using WLS. After the second fit he would again examine residuals to see if the WLS assumptions were valid. Maximum likelihood principles may be used to obtain more appropriate line-fitting techniques when errors are not normal. For example, LAD is better than OLS if the errors have a Laplace (double-exponential) probability distribution (Troutman, 1985a, 1985b).

In general, statistical line-fitting amounts to building a model for both the overall trend in the data (the straight line) *and* the structure of the errors (Troutman, 1985a, 1985b). This model-building procedure is an iterative process that alternates between fitting a line based on given assumptions and examining residuals to verify the assumptions. This process requires closer examination of the data and results than does simple application of a line-fitting formula. The familiarity that one gains with the data set in going through this process is one considerable advantage in approaching the problem in this manner.

DISCUSSION AND RECOMMENDATIONS

Eye-fitted lines

The "eye" method may be as accurate as some data warrant. Also, on rare occasions, investigators may know that some points on the graph are more reliable than others, and they can take this into consideration in drawing the line. With most data, however, different people usually do not draw the same line by eye. In addition to this subjectivity, no estimate of accuracy or precision of the line's slope and intercept are provided. For these reasons, fitting by eye is becoming less common and less satisfactory.

Averages

This method is simple, is more rigorous than drawing a line by eye, and has been rather popular in the past. However, the method can give different lines depending on how the data are grouped, the equation cannot be inverted (rearranged) or otherwise manipulated, and the equation is not as good from a statistical point of view as are the more rigorous methods discussed below.

Ordinary least squares

Because OLS provides a good estimator of the regression line under general conditions, it is often the best method to use when prediction is the goal. Another positive feature of OLS is that a mathematical equivalence can be established with any changes in scale, that is, OLS is consistent with scale changes. If a method is consistent with scale changes, then, when all of the X data are multiplied by g and all Y data are multiplied by h, we get a regression slope b' of these new variables of

$$b' = \frac{h}{g} b \tag{22}$$

(Hirsch and Gilroy, 1984). This relation holds when b' and b are determined by least squares.

There are, however, some disadvantages with OLS. First, it is based on assumptions that may not hold: (1) for each X, associated replicate measurements of Y are normally distributed and have a mean that falls on the "best-fit" line, (2) the different normal distributions for different values of X all have the same variance ("constant variance"), and (3) the errors or residuals for different values of X are independent of one another. If the errors are not normally distributed, the OLS line still has some optimal properties from a statistical point of view, but it may not be any better than straight lines obtained by other methods (Waugh, 1952, p. 314).

Second, the OLS equation for X on Y generally is not the same as a simple rearrangement of the equation for Y on X. Where values of X are predetermined by the experimenter and are measured without error, estimates of X from a known Y should be made from a rearrangement of the regression of Y on X (Griffiths, 1967, p. 449; Graybill, 1976, p. 277); where X instead is a random variable (values not being fixed in advance by the experimenter), the dependent variable in the regression should be the variable to be estimated, regardless of which variable governs the other (Winsor, 1946; Williams, 1983).

It is commonly believed that OLS requires a causal relation or physical dependence between variables and that the independent variable must be measurable without significant error; neither of these beliefs is true. If X does have error, but if prediction of Y is the purpose of the analysis, then OLS is still appropriate. If X is subject to error and if estimation of the "true" slope and intercept is the purpose of the analysis, then structural analysis probably is better.

OLS yields two different lines or relations (neither of which may represent the actual relation between the variables), depending on which variable is classified as dependent or independent in the regression (Figure 7.2). This is disturbing to physical scientists searching for an assumed fixed functional relation between the variables. From the viewpoints of both intuitive theoretical expectations and utility, they would much prefer a method that provides a unique line regardless of the axes to which the variables are assigned.

Weighted least squares

Weighted least squares is similar to OLS in that the regression relation (rather than the structural relation) is being estimated, and again this relation usually is of interest when prediction of a variable is the goal of the analysis. WLS is more suitable than OLS if the assumptions of constant variance and (or) unrelated errors are not obeyed. Weights are then assigned to reflect the fact that some measurements are less reliable than others. WLS has many of the same advantages and disadvantages of OLS except that the requirement of constant variance is relaxed. WLS is, however, more complicated.

Robust methods

Robust methods are similar to weighted least squares in that weighting is used in both techniques. However, the main purpose of WLS is to correct for unequal variances of errors at a fixed X. The weights usually are determined by considering the residuals at all X's as a group. Robust methods, in contrast, are intended mainly to correct for outliers. The weights usually are determined individually at each X, without being influenced by the residuals at other X's.

Robust methods are attractive from the viewpoint of adjusting for the effect of presumed outliers. However, rather than blindly applying a robust procedure, the researcher first should examine the data set carefully to identify potentially influential points (Cook and Weisberg, 1982). A possible explanation for outlying observations should be found; if an observation is believed to be truly in error or if, in retrospect, it should not have been included in the data set in the first place, it should be deleted from the fitting process altogether (Mackin, 1963, p. 139–140). (However, just because a point plots as an "outlier" is not a sufficient reason for throwing it out!) It is best to try to justify robust techniques, such as LAD, on the basis of analysis of residuals. Finally, calculations in robust regression are somewhat more complicated than with many other line-fitting methods.

Least normal squares

The LNS line has the intuitive appeal of being, by definition, the line that fits the data most closely, in that it minimizes the sum of squared perpendicular distances between the points and the line (Adcock, 1877). In addition, it is invertible (reversible). Thus, in contrast to OLS, a unique relation is obtained, regardless of which variable is chosen to be dependent (i.e., regardless of the axes to which the variables are assigned).

LNS does have the disadvantage that it is inconsistent with scale changes. This means that when the units of X or Y (or both) are changed, the new slope b' cannot be calculated from (22) but must instead be determined from a new fit. Thus, after rescaling, the new line does not have the same relation to the new pairs of (X, Y) values as the original line did to original pairs (Greenall, 1949, p. 40).

As pointed out before, LNS actually is a special case of structural analysis ($\lambda = 1$ in equation 8), if we choose to place our problem into a statistical framework. Thus, before applying the LNS technique, the researcher should justify the assumption that the magnitudes of the errors in both variables are close to being equal. (The errors here are reckoned in absolute terms, not percentage terms.) This assumption in many cases is not justifiable.

Reduced major axis

As with LNS, the RMA line is invertible so that the same equation can be used to solve for X or Y. Another advantage is that, as shown by (11), the slope of the RMA line is unaffected by the correlation coefficient, whereas the slopes of the OLS and LNS lines, (7) and (10), are. Also, the RMA method is invariant under scale changes; if the units of X and (or) Y were to be changed, the new slope could be calculated from the original slope using (22).

The RMA equation is not as good statistically as the appropriate OLS equation for predicting values of the dependent variable. However, according to Greenall (1949, p. 37), when $r > 0.7$, the standard deviation of error of estimate is increased by less than 8.5% by using the RMA line instead of the appropriate OLS line. The differences become less as the correlation improves. At the other extreme, the RMA technique gives a nonzero computed slope b (equation 11) even where there is no correlation ($r = 0$).

The RMA line roughly assumes that X^* and Y^* are measured with equal accuracy on a percentage basis (for example, in Figure 7.2, about 10% error in D and 10% error in Q). (This equality on a percentage basis is to be contrasted with the LNS assumption of equality of errors in absolute terms.) The assumption of equality on a percentage basis seems reasonable for many types of data. Greenall (1949, p. 35–36) and Kritskiy and Menkel (1968, p. 85) give further discussion. Thus, a good case can be made for using RMA when the goal is to find the underlying structural relation between two variables and when the assumption of roughly equivalent measurement accuracy on a percentage basis is acceptable.

Regression and structural analysis

Whether or not to place the line-fitting problem into a statistical framework is an option of the researcher. We feel that doing so is worthwhile, primarily because it forces the researcher to ask several important questions about the data and about the intended uses of the data. Is the fitted line to be used mainly for prediction? What is the magnitude of the errors in each of the variables? How much uncertainty is present in the computed line? Statistical analysis provides some *objective* guidelines for approaching these problems.

In a statistical analysis, the first problem is to decide whether the regression line or the structural line is more appropriate. If the regression line is to be used, the researcher must decide which variable should be the dependent variable. If the structural line is used, an estimate of the ratio (λ) of the error variances must be determined or assumed, in order to apply (8).

Many statistical techniques for analysis of the goodness of a fit are available. The process of analyzing the residuals after fitting a straight line may reveal peculiarities of the data that otherwise might go unnoticed. Likewise, in reporting the results of fitting a straight line, some statistical measures of the adequacy of the line should always be included.

REFERENCES

Adcock, R. J., 1877, Note on the method of least squares: *The Analyst,* v. 4, p. 183–184.

Aitchison, John, and Brown, J. A. C., 1957, *The lognormal distribution:* Cambridge Univ. Press, London, 176 p.

Chatterjee, Samprit, and Price, Bertram, 1977, *Regression analysis by example:* Wiley, New York, 228 p.

Cook, R. D., and Weisberg, Sanford, 1982, *Residuals and influence in regression:* Chapman and Hall, New York, 230 p.

Draper, N. R., and Smith, H., 1981, *Applied regression analysis:* Wiley, New York, 709 p.

Freund, R. J., and Littell, R. C., 1981, *SAS for linear models:* SAS Institute, Inc., Cary, N. C., 231 p.

Graybill, F. A., 1976, *Theory and application of the linear model:* Duxbury Press, North Scituate, Mass., 704 p.

Greenall, P. D., 1949, The concept of equivalent scores in similar tests: *Brit. Jour. Psych., Stat. Sec.,* v. 2, p. 30–40.

Griffiths, J. C., 1967, *Scientific method in analysis of sediments:* McGraw-Hill, New York, 508 p.

Hirsch, R. M., and Gilroy, E. J., 1984, Methods of fitting a straight line to data: Examples in water resources: *Water Resources Bull.,* v. 20, no. 5, p. 705–711.

Hogg, R. V., 1979, Statistical robustness: One view of its use in applications today: *The American Statistician,* v. 33, no. 3, p. 108–115.

Huber, P. J., 1973, Robust regression: Asymptotics, conjectures and Monte Carlo: *Ann. Stat.,* v. 1, no. 5, p. 799–821.

Huber, P. J., 1981, *Robust statistics:* Wiley, New York, 308 p.

Huynh, H., 1982, A comparison of four approaches to robust regression: *Psych. Bull.,* v. 92, no. 2, p. 505–512.

Imbrie, John, 1956, Biometrical methods in the study of invertebrate fossils: *Bull. Amer. Mus. Nat. Hist.,* v. 108, art. 2, p. 211–252.

Jansson, Margareta, 1985, A comparison of detransformed logarithmic regressions and power function regressions: *Geografiska Annaler,* v. 67A, no. 1–2, p. 61–70.

Jones, T. A., 1979, Fitting straight lines when both variables are subject to error, I., Maximum likelihood and least-squares estimation: *Math. Geology,* v. 11, no. 1, p. 1–25.

Kendall, M. G., and Stuart, Alan, 1967, *The advanced theory of statistics, Volume 2—Inference and relationship* 2nd ed.: Hafner, New York, 690 p.

Kritskiy, S. N., and Menkel, M. F., 1968, Some statistical methods in the analysis of hydrologic series: *Soviet Hydrology:* Selected Papers, no. 1, p. 80–98.

Lakshminarayanan, M. Y., and Gunst, R. F., 1984, Estimation of parameters in linear structural relationships: Sensitivity to the choice of the ratio of error variances: *Biometrika,* v. 71, no. 3, p. 569–573.

Mackin, J. H., 1963, Rational and empirical methods of investigation in geology, *in* Albritton, C. C., Jr. (ed.), *The fabric of geology:* Freeman, Cooper, Stanford, California, p. 135–163.

Madansky, Albert, 1959, The fitting of straight lines when both variables are subject to error: *Jour. Amer. Stat. Assoc.,* v. 54, p. 173–205.

Mark, D. M., and Church, Michael, 1977, On the misuse of regression in earth science: *Math. Geology,* v. 9, no. 1, p. 63–75.

Mejía, J. M., and Rodríguez-Iturbe, Ignacio, 1974, Correlation links between normal and log normal processes: *Water Resources Research,* v. 10, no. 4, p. 689–690.

Miller, D. M., 1984, Reducing transformation bias in curve fitting: *Amer. Stat.,* v. 38, no. 2, p. 124–126.

Mood, A. M., Graybill, F. A., and Boes, D. C., 1974, *Introduction to the theory of statistics* 3rd ed.: McGraw-Hill, New York, 564 p.

Osterkamp, W. R., McNellis, J. M., and Jordan, P. R., 1978, Guidelines for the use of structural versus regression analysis in geomorphic studies: *U.S. Geological Survey Water-Resources Investigations 78–135,* Lawrence, Kansas, 22 p.

Pearson, Karl, 1901, On lines and planes of closest fit to systems of points in space: *Philosophical Magazine,* 6th series, v. 2, no. 11, p. 559–572.

Till, Roger, 1973, The use of linear regression in geomorphology: *Area,* v. 5, no. 4, p. 303–308.

Troutman, B. M., 1982, An analysis of input errors in precipitation-runoff models using regression with errors in the independent variables: *Water Resources Research,* v. 18, no. 4, p. 947–964.

Troutman, B. M., 1983, Runoff prediction errors and bias in parameter estimation induced by spatial variability of precipitation: *Water Resources Research,* v. 19, no. 3, p. 791–810.

Troutman, B. M., 1985a, Errors and parameter estimation in precipitation-runoff modeling, 1. Theory: *Water Resources Research,* v. 21, no. 8, p. 1195–1213.

Troutman, B. M., 1985b, Errors and parameter estimation in precipitation-runoff modeling, 2. Case Study: *Water Resources Research,* v. 21, no. 8, p. 1214–1222.

Waugh, A. E., 1952, *Elements of statistical method:* McGraw-Hill, New York, 531 p.

Williams, G. P. 1983, Improper use of regression equations in earth sciences: *Geology,* v. 11, no. 4, p. 195–197 and discussion, 1984, v. 12, no. 2 p. 125–127.

Winsor, C. P., 1946, Which regression?: *Biometrics Bulletin,* v. 2, no. 6, p. 101–109.

Zucker, L. M., 1947, Evaluation of slopes and intercepts of straight lines: *Human Biology,* v. 19, p. 231–259.

Zucker, L. M., 1948, Discussion of La relation d'allometrie sa signification statistique et biologique, by Georges Teissier: *Biometrics,* v. 4, no. 1, p. 14–53.

8

Algebraic Manipulation of Equations of Best-Fit Straight Lines

Garnett P. Williams
and
Brent M. Troutman

The equation of a straight line fitted to data contains a deceptive equals sign. The equals sign is deceptive because most earth scientists assume that the equation can be manipulated algebraically, as with a purely mathematical relation. For example, investigators often rearrange a fitted equation $Y = a + bX$ into the form $X = (Y - a)/b$, in an effort to estimate X from Y. Unfortunately, such a rearrangement may not yield the same equation (i.e., the same constants) as a direct fit of X on Y (Williams, 1983). In this chapter we examine the behavior of five common algebraic operations on the equations of "best-fit" lines.

Three different line-fitting methods are examined: ordinary least squares (OLS), least normal squares (LNS), and the reduced major axis (RMA). OLS (regression of Y on X) produces the line having the minimum sum of squared vertical (parallel to Y axis) deviations between each data point and the line. LNS is a similar procedure, but the deviations between a point and the line are normal to the line rather than vertical. The RMA line is obtained by minimizing the sum of areas of right triangles between the points and the line, with the orthogonal sides of the triangles being parallel to the axes of the graph. These three techniques generally yield different lines for the same data. Imbrie (1956), Hirsch and Gilroy (1984), and

Troutman and Williams (Chapter 7) give fitting procedures and further explanation.

This analysis deals only with random variables. A random variable is one for which a value is not predetermined or chosen by the researcher but instead is characterized by a probability distribution.

TYPES OF ALGEBRAIC MANIPULATION

Manipulations explained here will be in terms of power laws ($Y = aX^b$), because many sets of geologic data plot as straight lines on log scales. However, power laws can be written in the same form as arithmetic straight lines simply by writing the equation in log form ($\log Y = \log a + b \log X$). Thus the examples illustrated here for power laws also apply to straight lines on arithmetic and semilog scales. All equations in the examples are "best-fit" lines to data. The five types of manipulation to be discussed are:

1. *Rearrangement.* This involves "inverting" a fitted equation $Y = aX^b$ to solve for X, so that $X = (Y/a)^{1/b}$ or $X = (1/a)^{1/b} Y^{1/b}$.

2. *Equating two equations.* Two different fitted equations having the same dependent variable presumably might be equated to solve for one independent variable from a knowledge of the other. For example, if $Y = aX^b$ and $Y = cZ^f$ we might assume that $cZ^f = aX^b$ or $Z = (a/c)^{1/f} X^{b/f}$.

3. *Substitution of one equation into another.* From $Y = aX^b$ and $X = cZ^f$, substitution of cZ^f into the place of X in the former equation gives $Y = a(cZ^f)^b$ or $Y = (ac^b)Z^{fb}$. In practice, substitution commonly is done in two separate steps rather than directly, but the end result is the same. That is, investigators usually first compute X from $X = cZ^f$ and then, as a second step, insert this computed X value into $Y = aX^b$ to calculate Y. This is simply a slightly more laborious way of arriving at $Y = (ac^b)Z^{fb}$.

4. *Multiplication of two equations.* If $Y = aX^b$ and $Z = cX^f$, multiplication of Y times Z gives $YZ = aX^b \cdot cX^f$ or $YZ = acX^{b+f}$.

5. *Division of one equation by another.* Using the same basic equations as in multiplication, division of Y by Z yields $(Y/Z) = (aX^b/cX^f) = (a/c)X^{b-f}$.

The key question in each of these five cases is whether the indicated manipulation gives an equation identical to that obtained by a direct fit of the variables.

THEORETICAL RESULTS

A theoretical analysis is sufficient to show which manipulations are valid for each of the three methods of fitting straight lines. Such a mathematical analysis is given in Appendix 1 and is summarized here in Table 8.1.

TABLE 8.1 Summary of Mathematical Validity of Algebraic Manipulation of Power Laws Fitted by OLS, LNS, and RMA Methods

Type of manipulation	Ordinary least squares	Least normal squares	Reduced major axis
Rearrangement	No[a]	Yes[a]	Yes
Equating	No	No	Yes[b]
Substitution	No	No	Yes[b]
Multiplication (addition)	Yes	No	No
Division (subtraction)	Yes	No	No

[a]Yes means the manipulation is valid; no means the manipulation is invalid or is valid only in some cases.
[b]Calculated slopes will have the correct magnitude but the signs may be wrong; this is not a problem if the correlation is high.

As Table 8.1 shows, rearrangement is invalid for any equation fitted by ordinary least squares. Equating and substitution are invalid for both OLS and LNS equations. Yet, examples of these three types of manipulation (especially rearrangement of an OLS equation) are common in the earth-sciences literature.

ERROR BOUNDS

Where a manipulation is not valid, we ideally would like to be able to determine from theoretical considerations what degree of error might result from using the manipulation. If this error is small in most cases, the manipulation, although not exactly valid, probably could be applied safely.

In many instances, simple and easy-to-apply error bounds may not be obtainable. This is particularly true for LNS lines, as the complexity of the equation for the slope makes theoretical results difficult to derive. However, we can obtain theoretical error estimates for a few special cases (Appendix 2). Typical errors associated with selected field data are determined in the next section.

EXAMPLES USING EMPIRICAL DATA

Hydraulic-geometry data of rivers were used with the invalid manipulations of Table 8.1 to get an approximate idea of typical magnitudes of errors. Such data can be used with all five types of manipulation and were available. In hydraulic-geometry relations, the water-surface width W and mean flow depth, D, at a stream cross section change in proportion to some power of water discharge, Q:

$$W = aQ^b \tag{1}$$

and

$$D = cQ^f \tag{2}$$

Measured width, depth, and discharge data were obtained from U.S. Geological Survey files for 10 streamflow-gaging stations in the United States. Sites were selected primarily to provide a variety of data sets in regard to slopes of fitted lines and of correlations between variables. Slopes of fitted straight lines (exponents in equations 1 and 2) ranged from 0.0012 to 0.60. Values of correlation coefficient for the log form of (1) and (2) ranged from 0.36 to 0.99; most were above 0.90. The number of data points per station ranged from 12 to 39. Although less than ideal from some statistical viewpoints, such sample sizes are typical of many geologic data sets.

Each of the five types of manipulation produces a simple power-law equation that can be compared with a corresponding directly fitted equation of the same variables. Differences between the two lines (manipulated versus directly fitted) were evaluated as follows. First, percent error for the value of Y indicated by the manipulated line was computed for each extreme of the range of X values of each data set. (X in this sense is the variable that is independent after manipulation of the equation.) At the smallest and again at the largest value of X, the percent error in Y is simply $100\,(|Y_d - Y_m|/Y_d)$, where Y_d is the value of Y indicated by the directly fitted line and Y_m is that of the manipulated line, at the given X. Maximum percent error between lines occurs at one of these extremes. Minimum possible percent error for all invalid manipulations always was zero; within the range of X values for each data set, the manipulated and directly fitted lines intersected approximately at the point $\overline{X}, \overline{Y}$, so that at this point of intersection both equations predict the same Y-value.

A mean squared error E (weighted on the basis of an assumed normal distribution of values of the new independent variable) between two straight lines also was computed from

$$E = [a_d - a_m + (b_d - b_m)\mu]^2 + (b_d - b_m)^2\sigma^2 \qquad (3)$$

where a_d is the intercept of the directly fitted line, a_m is that of the manipulated line, b_d and b_m are the slopes of those lines, and μ and σ^2 are the mean and variance of the newly defined independent variable. With a_d, a_m, μ, and σ^2 in natural log units, a standard error in percent was computed as $100(e^E - 1)^{0.5}$, where $e = 2.718$. This percentage combines the errors in slope and intercept and reflects an overall weighted mean error between the two lines; it does not represent the error in the prediction of the dependent variable at any particular value of the independent variable.

For the data used here, the number of decimal places maintained in the computations can produce a difference of as much as about 3% in the calculated error percentages. Thus there is little or no significant difference between percentages that are within that range of one another. By the same token, percentages of less than about 3% may not be significantly different from zero. Also, errors as large as about 5% can occur just from a difference of 0.01 in either the slope or intercept.

The first type of manipulation, rearrangement, involved inverting $D = cQ^f$ (equation 2) to the form $Q = (D/c)^{1/f} = (1/c)^{1/f}D^{1/f}$. As Table 8.1 indicates, this operation gives the same equation as a direct fit of Q on D for LNS and RMA lines but not for OLS lines. Using the directly fitted OLS line (Q on D) as the correct relation and comparing Q-values indicated by the two lines, the maximum error in rearranging an OLS equation for the 10 field cases ranged from 4 to 880%. The weighted mean error between lines ranged from 2 to 106%.

Rearrangement of a multiple regression (OLS) equation (solving for one of two or more independent variables) also is commonly done. The errors in such manipulation can be at least as large as, and possibly greater than, those associated with the simpler equations (Williams, 1983).

The second type of manipulation, equating, involved the use of an observed power law between W and D, namely, $W = kD^n$. We have, therefore, two fitted relations: $W = aQ^b$ in (1) and also $W = kD^n$. Equating the right-hand sides of these two expressions for W results in $kD^n = aQ^b$ or $D = (a/k)^{1/n}Q^{b/n}$. According to Table 8.1, this operation produces the same constants as the direct fit of D on Q in (2) only if the fitted lines are the reduced major axis. Equating OLS relations resulted in maximum between-line errors of 2 to 46% and weighted mean errors of 2 to 30%, for the 10 field cases. For LNS relations, the maximum error between lines ranged from 0 to 28%, and the weighted mean error varied from 0 to 11%.

From two fitted relations $W = aQ^b$ and $Q = mD^z$, substitution of the latter for Q in the former relation gives W as a function of D: $W = aQ^b = a(mD^z)^b$, or $W = (am^b)D^{zb}$. This operation produces the same coefficient and exponent as would be obtained by a direct fit of W on D only for RMA lines (Table 8.1). Performing this operation on OLS equations for the 10 field cases yielded maximum between-line errors of 0 to 34% and weighted mean errors of 0 to 20%; for LNS lines, these errors ranged from 0 to 70% and from 0 to 27%, respectively.

The water-surface width W times the mean flow depth D by definition gives cross-sectional flow area, and this also bears a power relation to Q. A direct fit of WD on Q gives the same equation as that obtained by taking (1) and (2) and computing $WD = aQ^b cQ^f = acQ^{b+f}$ only if the lines are fitted by OLS (Table 8.1). Lines fitted by LNS to the 10 field cases had maximum between-line errors of 0 to 9% and weighted mean errors of 0 to 4%; for RMA lines, these errors were 0 to 22% and 0 to 12%, respectively.

The ratio W/D on log paper also plots as a straight line with Q. Using (1) and (2), we can presumably write this ratio as $W/D = aQ^b/cQ^f = (a/c)Q^{b-f}$. Such manipulation consistently gives the same equation as a direct fit of W/D on Q only for OLS lines (Table 8.1). With lines fitted by LNS and RMA, division of equations may or may not yield the same constants as does the direct fit. For the 10 streamflow-gaging stations, errors between lines fitted by LNS generally were negligible, and the discussion of error bounds (Appendix 2) makes allowances for this. (There are conditions under which LNS lines can't be divided reliably.) The between-line maximum errors associated with RMA-fitted lines ranged from negligible to 94%, and weighted mean errors ranged from negligible to 34%.

COMPOUND MANIPULATIONS

Compound manipulations involve two or more of the five basic types of manipulations discussed above. If algebraic manipulation is invalid for one of the five basic methods, it will be similarly invalid for a compound manipulation that includes such a method. Here are two examples.

The first consists of both rearranging and equating. Consider (1) and (2), namely, $W = aQ^b$ and $D = cQ^f$. The presence of Q in both expressions leads one to assume that W can be related to D. Rearranging each equation gives $Q = (W/a)^{1/b}$ and $Q = (D/c)^{1/f}$. A second manipulation, equating, gives $(W/a)^{1/b} = (D/c)^{1/f}$ or $W = (a/c^{b/f})D^{b/f}$. Such manipulations would be invalid if the original lines were fitted either by OLS or by LNS (Table 8.1). A common situation in which these manipulations are done in paleohydrology involves theoretical and empirical expressions for the shear stress needed to entrain a sediment particle on a streambed. These expressions commonly are rearranged and equated in different ways to solve for one component variable of shear stress.

A second example consists of multiple substitution and involves the Gauckler–Manning formula of river hydraulics. This gives mean flow velocity V as a function of mean water depth D, channel slope (or energy gradient) S, and a resistance coefficient, N:

$$V = D^{2/3}S^{1/2}/N \tag{4}$$

Empirically, it is known that each of the three independent variables can be written as a separate power function of water discharge, Q: $D = cQ^f$ (equation 2), $S = mQ^z$, and $N = kQ^n$. In practice, values of the coefficients and exponents in these three equations might be available from "best-fit" relations to actual data, and the researcher might want a relation between V and Q. Substituting each of the three power functions into the place of the respective independent variable in (4) at first glance would set up this relation. Thus, $V = (cQ^f)^{2/3}(mQ^z)^{1/2}/kQ^n$, or $V = (c^{2/3}m^{1/2}/k)Q^{(2f/3) + (z/2) - n}$. Again, the results shown above indicate this relation would be valid, that is, would give the same coefficient and exponent as a direct fit of V on Q, only if all lines were fitted by RMA.

Within the earth sciences, compound manipulations are particularly common in paleohydrology, river hydraulics, sediment transport, and probably in many other disciplines. The operations usually include various combinations of rearranging, equating, and substitution. Instances where authors have "combined" several equations into a new equation should be inspected carefully in regard to (a) the manner in which the constants in the original equations were determined (not uncommonly, an author does not even explain how he or she fitted a line to data) and (b) whether the method of "combining" the equations is valid.

CONCLUSIONS

Equations determined by the OLS, LNS, and RMA techniques can be manipulated safely in some ways but not in others. OLS equations can be multiplied and divided, but they can't safely be rearranged, equated, or substituted. Unfortunately, these last three operations are particularly popular, and so is the OLS method of fitting a line. LNS equations can be rearranged, multiplied, and divided but can't always be equated to, or substituted in, one another. RMA equations can be rearranged, equated, and substituted but not necessarily multiplied or divided. Where a particular type of manipulation is not necessarily valid, the maximum possible error may be estimated in some cases. This estimate may indicate whether the manipulation can safely be applied anyway.

REFERENCES

Hirsch, R. M., and Gilroy, E. J., 1984, Methods of fitting a straight line to data: Examples in water resources: *Water Resources Bulletin,* v. 20, no. 5, p. 705–711.

Imbrie, John, 1956, Biometrical methods in the study of invertebrate fossils: *Bull. Amer. Mus. Nat. Hist.,* v. 108, art. 2, p. 211–252.

Kaplan, Wilfred, 1952, *Advanced calculus:* Addison–Wesley, Reading, Mass., 679 p.

Karlinger, M. R., and Troutman, B. M., 1985, Error bounds in cascading regressions: *Jour. Internat. Assoc. Math. Geol.,* v. 17, no. 3, p. 287–295.

Williams, G. P., 1983, Improper use of regression equations in earth sciences: *Geology,* v. 11, p. 195–197 and v. 12, p. 125–127.

APPENDIX 1: THEORY

Assume that we have a sample of N measurements on three variables (X, Y, Z); all calculations are based on this single sample. The sample means, variances, and covariance for X and Y are computed in the standard manner:

$$\bar{X} = \frac{1}{N} \sum_{i=1}^{N} X_i \tag{5}$$

$$\bar{Y} = \frac{1}{N} \sum_{i=1}^{N} Y_i \tag{6}$$

$$S_x^2 = \frac{1}{N} \sum_{i=1}^{N} (X_i - \bar{X})^2 \tag{7}$$

$$S_y^2 = \frac{1}{N} \sum_{i=1}^{N} (Y_i - \bar{Y})^2 \tag{8}$$

and

$$S_{xy} = \frac{1}{N} \sum_{i=1}^{N} (X_i - \bar{X})(Y_i - \bar{Y}) \tag{9}$$

where N is the sample size, Σ means summation, $\bar{X}$ is the mean and S_x^2 is the variance of the sample of X-values, $\bar{Y}$ and S_y^2 are the mean and variance of the Y values, S_{xy} is the covariance between X and Y, and the subscripts i refer to individual observations of the associated variable. Statistics S_{xz}, S_{yz}, $\bar{Z}$, and S_z^2 are computed in a similar manner. Also, we let $W = Y + Z$ and $U = Y - Z$, and compute S_{xw}, S_{xu}, S_w^2, and S_u^2. Then the following relations hold:

$$S_{xw} = S_{xy} + S_{xz} \tag{10}$$

$$S_{xu} = S_{xy} - S_{xz} \tag{11}$$

$$S_w^2 = S_y^2 + S_z^2 + 2S_{yz} \tag{12}$$

and

$$S_u^2 = S_y^2 + S_z^2 - 2S_{yz} \tag{13}$$

Formulas for the best-fit lines between any two variables are given in Troutman and Williams (see Chapter 7). Let H and K each stand for X, Y, Z, U, or W. Equations for the slope b of the best-fit lines with H as the dependent variable and K as the independent variable are:

$$\text{OLS:} \quad b_{hk} = \frac{S_{hk}}{S_k^2} \tag{14}$$

$$\text{LNS:} \quad b_{hk} = \frac{S_h^2 - S_k^2 + [(S_h^2 - S_k^2)^2 + 4S_{hk}^2]^{1/2}}{2S_{hk}} \tag{15}$$

$$\text{RMA:} \quad b_{hk} = \text{sign} \, (S_{hk}) \frac{S_h}{S_k} \tag{16}$$

The first subscript on the computed slope b indicates the dependent variable, and the second subscript indicates the independent variable. For example, b_{xy} would denote a slope computed with X as the dependent var-

TABLE 8.2 Summary of Slopes of Lines Obtained by Direct Fit and by Manipulation

Type of manipulation	Slope of direct-fit line	Slope of manipulated line(s)
Rearrangement	b_{xy}	$1/b_{yx}$
Equating	b_{zx}	b_{yx}/b_{yz}
Substitution	b_{yz}	$b_{yx}b_{xz}$
Multiplication (addition)	b_{wx}	$b_{yx} + b_{zx}$
Division (subtraction)	b_{ux}	$b_{yz} - b_{zx}$

iable and Y as the independent variable, and b_{yx} would be the slope computed with the roles of the variables reversed. Intercepts are determined from the computed slope and from the requirement that all three lines pass through $(\bar{H}, \bar{K})$.

Table 8.1 shows which algebraic manipulations are valid for the three methods of fitting straight lines. ("Valid" here means that a manipulation produces the same constants as does a direct fit of the variables.) Because all lines pass through the point with sample means of the two variables as coordinates, a given operation is valid if and only if it works for the slopes of the lines. Thus, we examine only slopes in what follows. Also, because multiplication is equivalent to addition and division to subtraction, using log-transformed variables, we restrict our attention to addition and subtraction in this section.

Table 8.2 summarizes the slopes for lines obtained by a direct fit and by manipulation. Regardless of the line-fitting technique, equality of the given slopes must hold for the manipulation to be valid. Consider, for example, the "rearrangement" row in Table 8.2. A direct-fit line computed with X as the dependent variable and Y as the independent variable has slope denoted by b_{xy}, according to our convention. The equation of the direct-fit line is $X = a + b_{xy}Y$, where a is the intercept. Now, the line computed with Y as the dependent variable has slope b_{yx}, and the equation of this line is $Y = a' + b_{yx}X$, where a' is the intercept. Rearrangement of this equation gives $X = (-a'/b_{yx}) + (1/b_{yx})Y$; thus the slope of the manipulated line is $1/b_{yx}$, as shown in Table 8.2. For the manipulation to be valid, we must have $b_{xy} = 1/b_{yx}$. Four cases of manipulation, shown in the next paragraph, are always valid (Table 8.1). We first prove the validity for these cases.

(1) Multiplication, OLS

For the multiplication of two OLS lines (or sum of log-transformed values) to be valid, the slope given by such a manipulation ($b_{yx} + b_{zx}$) must equal b_{wz}, the direct-fit slope (see multiplication row in Table 8.2). Using (14) as a basis, then inserting the relation $S_{wz} = S_{yx} + S_{zx}$ (equation 10 with roles of variables reversed), and finally using (14) again, we have

$$b_{wx} = \frac{S_{wx}}{S_x^2} = \frac{S_{yx} + S_{zx}}{S_x^2} = \frac{S_{yx}}{S_x^2} + \frac{S_{zx}}{S_x^2} = b_{yx} + b_{zx} \tag{17}$$

Equation 17 shows that, for this case, the direct-fit slope does indeed equal the manipulated slope.

(2) Division, OLS

This is shown in the same way as multiplication, OLS (above), except that we use $S_{ux} = S_{yx} - S_{zx}$.

(3) Rearrangement, LNS

According to Table 8.2, rearrangement of a fitted line of slope b_{xy} must yield a new slope equal to $1/b_{yx}$, for such a manipulation to be valid. Equation 15 with variables X (dependent) and Y (independent) is:

$$b_{xy} = \frac{S_y^2 - S_x^2 + [(S_y^2 - S_x^2)^2 + 4S_{xy}^2]^{1/2}}{2S_{xy}}$$

This yields

$$b_{xy} = \frac{(S_y^2 - S_x^2)^2 - [(S_y^2 - S_x^2)^2 + 4S_{xy}^2]}{2S_{xy}\{S_y^2 - S_x^2 - [(S_y^2 - S_x^2)^2 + 4S_{xy}^2]^{1/2}\}}$$

when both numerator and denominator are multiplied by the quantity inside the braces { }. The numerator's terms reduce to $-4S_{xy}^2$, so that

$$b_{xy} = \frac{-4S_{xy}^2}{2S_{xy}\{S_y^2 - S_x^2 - [(S_y^2 - S_x^2)^2 + 4S_{xy}^2]^{1/2}\}}$$

$$= \frac{2S_{xy}}{S_x^2 - S_y^2 + [(S_x^2 - S_y^2)^2 + 4S_{xy}^2]^{1/2}} = \frac{1}{b_{yx}} \tag{18}$$

(4) Rearrangement, RMA

From equation (16), we have:

$$b_{xy} = \text{sign}\,(S_{xy})\frac{S_y}{S_x} = \frac{1}{\text{sign}\,(S_{xy})\dfrac{S_x}{S_y}} = \frac{1}{b_{yx}} \tag{19}$$

To show that a manipulation is not valid, it is sufficient to exhibit a counterexample, or a particular combination of values of the statistics

TABLE 8.3　Calculated Values for Slopes of OLS, LNS, and RMA Lines Using Selected Sample Statistics

Calculated slope	Slope of ordinary least squares line	Slope of least normal squares line	Slope of reduced major axis
b_{xy}	0.25	0.41	0.71
b_{yx}	0.50	2.41	1.41
b_{xz}	0.50	0.98	0.58
b_{zx}	1.50	1.87	1.73
b_{yz}	0.17	0.41	0.82
b_{wx}	2.00	2.85	2.45
b_{ux}	−1.00	−3.30	−2.00

(S_x^2, S_y^2, etc.), for which the manipulation will not work. Consider the following set of sample statistics:

$$S_x^2 = 1.0, \qquad S_y^2 = 2.0, \qquad S_z^2 = 3.0$$

$$S_{xy} = 0.5, \qquad S_{xz} = 1.5, \qquad S_{yz} = 0.5$$

(There is a set of data that will lead to such statistics.) Calculated slopes for the three line-fitting methods are shown in Table 8.3. Substitution of these calculated slopes into Table 8.2 reveals that the manipulations shown as invalid in Table 8.1 do not work for this data set. For example, $b_{xy} = 0.25$ and $1/b_{yx} = 2.00$ for ordinary least squares. Therefore, rearrangement is not valid. Equating and substitution do work for RMA lines here. An instance for which equating and substituting will not work for RMA lines because of a sign error is obtained by taking $S_{yz} = -0.1$ instead of 0.5.

APPENDIX 2: ERROR ESTIMATES

This appendix presents theoretical estimates of the error associated with certain invalid manipulations. Errors resulting from substitution of OLS lines are discussed in detail by Karlinger and Troutman (1985). The situation they consider is as follows: One has pairs of observations (X, Y) and (X, Z) with which b_{yx} and b_{xz} may be computed, but one does not have paired measurements on (Y, Z) that could be used to obtain a direct OLS fit of Y on Z. This is one common situation for which this type of algebraic manipulation would be useful. They show that an upper bound for the difference in slopes (see Table 8.2) is

$$|b_{yz} - b_{yx}b_{xz}| < \left(1 - \frac{S_{yx}^2}{S_y^2 S_x^2}\right)\left(1 - \frac{S_{xz}^2}{S_x^2 S_z^2}\right)\frac{S_y}{S_z}$$

This bound also will hold under the assumption (made in this chapter) that all computations are based on a single sample of measurements on (X, Y, Z).

Another useful bound may be derived for addition of RMA lines. Assume that $S_{xy} \geq 0$ and $S_{xz} \geq 0$. This implies that $S_{xw} \geq 0$, from (10). Taking (16) and substituting (12) into the numerator gives

$$
\begin{aligned}
b_{wz} = \frac{S_w}{S_x} &= S_x^{-1}[S_y^2 + S_z^2 + 2S_{yz}]^{1/2} \\
&= S_x^{-1}[(S_y + S_z)^2 - 2(S_yS_z - S_{yz})]^{1/2} \\
&= S_x^{-1}(S_y + S_z)\left[1 - \frac{2(S_yS_z - S_{yz})}{(S_y + S_z)^2}\right]^{1/2} \\
&= (b_{yx} + b_{zx})\left[1 - \frac{2(S_yS_z - S_{yz})}{(S_y + S_z)^2}\right]^{1/2}
\end{aligned}
\tag{20}
$$

Using Taylor's theorem with remainder (Kaplan, 1952, p. 358), (20) leads to

$$
\frac{|b_{wx} - (b_{yx} + b_{zx})|}{b_{yx} + b_{zx}} \leq \frac{S_yS_z - S_{yz}}{(S_y + S_z)(S_y^2 + S_z^2 + 2S_{yz})^{1/2}} = C
\tag{21}
$$

which is an upper bound on the difference in slopes for a direct fit and a fit obtained by addition, expressed as a fraction of the slope obtained by addition. In other words, the percent error in the slope is no greater than $100C\%$.

Also, for $S_{yz} \geq 0$,

$$
C \leq \frac{S_yS_z}{S_y^2 + S_z^2} \leq 0.50
\tag{22}
$$

Thus the relative error is no more than 50%.

Analogous results may be obtained for the cases when either or both of S_{xy} and S_{xz} are negative, but care needs to be taken to keep track of the signs on the slopes. Also, subtraction of RMA lines may be treated similarly.

Bounds for multiplication and division (addition and subtraction under log transform) of LNS lines may be obtained under certain conditions. Consider, for example, the situation for which the variance of the dependent variable is small compared to that of the independent variable. The LNS slope b_{hk} in (15) may be expressed as a function of S_h/S_k and the correlation $r_{hk} = S_{hk}/(S_hS_k)$. Taylor's theorem with remainder (Kaplan, 1952, p. 358) may be applied to approximate b_{hk} for small S_h/S_k. We expand the expression for b_{hk} in (15) around the point $S_h/S_k = 0$ to obtain

$$
b_{hk} = r_{hk}(S_h/S_k) + R(S_h/S_k)^3
\tag{23}
$$

r_{hk} is the sample correlation coefficient between H and K. R obeys

$$|R| \leq g_{hk} \tag{24}$$

where g_{hk} is defined to be

$$g_{hk} = 1/[8r_{hk}^2(1 - r_{hk})^2] \tag{25}$$

Now, $r_{hk}(S_h/S_k)$ in (23) is another expression for the OLS slope in equation (14); see Troutman and Williams (Chapter 7). Therefore, (23) may be written as

$$b_{hk}(\text{LNS}) = b_{hk}(\text{OLS}) + R(S_h/S_k)^3 \tag{26}$$

or,

$$b_{hk}(\text{LNS}) - b_{hk}(\text{OLS}) = R(S_h/S_k)^3 \tag{27}$$

Using (24) leads to

$$|b_{hk}(\text{LNS}) - b_{hk}(\text{OLS})| \leq g_{hk}(S_h/S_k)^3 \tag{28}$$

This inequality shows that the LNS and OLS slopes should be close to each other provided S_h/S_k is small and g_{hk} is small.

Multiplication and division are valid for OLS (Table 8.1), or

$$b_{wx}(\text{OLS}) = b_{yx}(\text{OLS}) + b_{zx}(\text{OLS}) \tag{29}$$

We would expect the same equation to be nearly true for LNS, provided the bound in (28) is small. A bound for the error in multiplying LNS lines may be obtained as follows:

$$|b_{wx}(\text{LNS}) - [b_{yx}(\text{LNS}) + b_{zx}(\text{LNS})]|$$

$$= |b_{wx}(\text{LNS}) - b_{wx}(\text{OLS}) - \{[b_{yx}(\text{LNS}) - b_{yx}(\text{OLS})] + [b_{zx}(\text{LNS})$$

$$- b_{zx}(\text{OLS})]\}|$$

$$\leq |b_{wx}(\text{LNS}) - b_{wx}(\text{OLS})| + |b_{yx}(\text{LNS}) - b_{yx}(\text{OLS})| + |b_{zx}(\text{LNS})$$

$$- b_{zx}(\text{OLS})|$$

$$\leq g_{wx}(S_w/S_x)^3 + g_{yx}(S_y/S_x)^3 + g_{zx}(S_z/S_x)^3$$

The first equality follows from (29), the first inequality follows from the relation $|a + b| \leq |a| + |b|$, and the last inequality follows from (28). The same upper bound will work for division if u replaces w and $b_{yx} - b_{zx}$ replaces $b_{yx} + b_{zx}$. Generally, if S_w, S_y and S_z are small compared to S_x, multiplication and division of LNS lines should lead to small errors.

9

The Analysis of Bivariate Association

James A. Harrell

Data suitable for the analysis of bivariate association consist of samples with paired observations on two variables (e.g., X and Y). Such data are conveniently displayed in a scatter plot (Figure 9.1). The objective of association analysis is to measure the extent of linear and curvilinear covariation between two variables. Association analysis provides an answer to the basic question of "how strong is the tendency for Y to increase or decrease as X increases?" A statistic known as the correlation coefficient is used to describe the extent of covariation and, hence, the strength of bivariate association.

Association analysis is especially useful in the exploratory survey of multivariate data sets (i.e., multiple variables measured on a suite of samples). Such data sets are common in geology and the interrelationships between pairs of variables are normally of interest. A convenient way of summarizing these interrelationships is in a correlation matrix, in which the correlation coefficients for all possible pairs of variables are tabulated. The observed bivariate correlations may be put to a variety of uses: they may (1) help to confirm or refute a hypothesis (i.e., an expected relationship), or even lead to the formulation of new hypotheses; (2) suggest predictive relationships among variables by indicating which variables can be used to predict the values of others; (3) help to identify variables that can be used in subsequent investigations as surrogates for other variables that are too expensive or difficult to measure; and (4) be used as input to other statistical techniques such as R-mode cluster and principal components analyses.

The observed strength of a bivariate association has one of three origins.

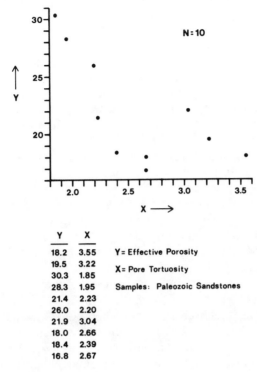

Y	X	
18.2	3.55	Y = Effective Porosity
19.5	3.22	X = Pore Tortuosity
30.3	1.85	
28.3	1.95	Samples: Paleozoic Sandstones
21.4	2.23	
26.0	2.20	
21.9	3.04	
18.0	2.66	
18.4	2.39	
16.8	2.67	

Figure 9.1 Scatter plot for a typical bivariate data set (Harrell, 1983).

It can be the result of a cause-and-effect relationship where, for example, Y takes on values that are a response to changes in the value of X. In other cases, X and Y covary because both have a mutual dependence on (i.e., cause-and-effect relationship with) some other variable(s). Finally, the relationship between X and Y can be entirely spurious if it arises from the chance covariation between two otherwise unrelated variables. Spurious relationships can, in many cases, be identified from a test of statistical significance. The distinction between cause-and-effect and mutual dependence relationships can only be made through a properly designed experiment.

TYPES OF BIVARIATE ASSOCIATION

There are two types of bivariate association: natural and experimental. In natural associations both X and Y are subject to random error. The observed values for each variable come from a population of values. In principle, this population is randomly sampled by an investigator who has no manipulative control over the values of the variables but simply accepts them as they occur in nature. Because the samples reflect the inherent variation within the population, it is said that the observed values contain random error. A second and generally smaller source of random error is the reproducibility of the analytical measurements made on the samples. An example of a natural association would be the covariation of bedload discharge and mean flow velocity in a stream. Another example is the relationship between sandstone porosity and tortuosity depicted in Figure 9.1.

In experimental associations Y is subject to random measurement and sampling errors as above whereas X contains only, generally negligible, measurement error. The values of X are largely without error because they are manipulated and exactingly set as required by an experimental design. An example of an experimental association would be the variation in bedload discharge in a laboratory flume for set mean flow velocities.

Correlation coefficients can be calculated for both natural and experimental associations. The distinction between the two types of associations is important, however, because each requires a different interpretation of the correlation coefficient. The correlation coefficient for natural associations is an estimate of the true covariation between X and Y in the sampled population as well as an empirical measure of the strength of the bivariate association. The sampled population ideally represents the set of all possible paired X, Y observations in the natural environment. For experimental associations, the correlation coefficient is simply a descriptive statistic that indicates the extent of covariation between X and Y under experimental conditions. In the latter case, the correlation coefficient may change with the experimental conditions and so is not an estimator of the X, Y correlation in the natural population. One can, of course, speak in terms of a population of X and Y values in an experimental situation. The population consists of a set of X values defined by an investigator and the corresponding observed Y values. For each X value there exists a distribution of Y values. Each Y distribution is, in effect, randomly sampled by the investigator to produce a paired X, Y observation from the experimental X, Y population.

GEOMETRY OF BIVARIATE TRENDS

The point swarm in a bivariate scatter plot may exhibit a linear or curvilinear trend, or no trend at all. The point swarm in Figure 9.1, for example, appears to have either a linear or slight, concave-up curvilinear trend.

Figure 9.2 illustrates the spectrum of generalized geometries for bivariate trends. At the far left of the spectrum is the straight line, the simplest type of bivariate trend. Immediately to the right are simple curves. These

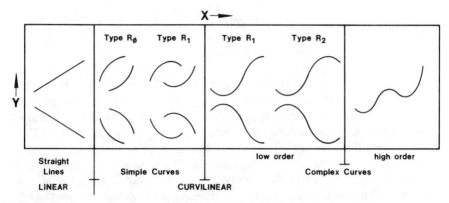

Figure 9.2   Spectrum of generalized geometries of bivariate trends.

curved lines have one bend or inflection. Type R_0 curves have no reversal in slope whereas type R_1 curves do exhibit a slope reversal. Complex curves are curved lines with two (low order) or more (high order) inflections. Among the low order curves, type R_1 has only one slope reversal and type R_2 has two.

It seems likely that the vast majority of "real" bivariate trends are of the linear and type R_0, simple curvilinear varieties. The slope reversals that characterize other curve varieties require relationships between pairs of variables that seem improbable. One variable should generally increase or decrease in some fashion with another, but it seldom makes sense that it should do both. When such trends do occur in sample data, they are probably the result of improper or inadequate sampling, or a lack of a real relationship (i.e., the trend is spurious). There are, of course, those bonafide but infrequent situations where real slope reversals in simple curvilinear trends do occur. There are also those occasions, such as in the case of time and distance series, where real complex curvilinear trends are expected.

The bulk of the statistical literature on bivariate associations is largely concerned with linear trends. The possibility of curvilinear trends is often ignored. This practice, in large part, reflects a widespread but erroneous belief that bivariate relationships are fundamentally linear.

CORRELATION AND REGRESSION ANALYSES

The degree of association between pairs of variables is traditionally investigated through either correlation analysis or regression analysis. Correlation analysis, as usually practiced by statisticians, involves more than just characterizing the strength of an empirical relationship. It is ultimately concerned with estimating (via the correlation coefficient) the *correlation* between X and Y in a natural population and so is appropriately applied only to natural associations. Regression analysis, in contrast, was originally developed for experimental associations and populations.

The goal of regression analysis is to derive a prediction equation that describes the functional relationship between X and Y (i.e., $\hat{Y} = f[X]$ where $\hat{Y}$ is the predicted value of Y for a given value of X). The equation defines a line that best represents the trend of the point swarm in a scatter plot. One of the two variables is designated the response variable and the other the predictor variable. The symbols conventionally used to represent these variables are Y and X, respectively. These same symbols are also commonly used in correlation analysis but they have no functional connotations. The terminology used in regression analysis stems from the fact that the prediction equation uses X to predict Y, which supposedly varies in response to changes in X. As is appropriate for experimental associations, the values of Y are assumed to be subject to random error whereas the values of X are considered to be exact. The practical importance of this is that the prediction equation is derived so that it minimizes the sum of the squared deviations ($\Sigma[Y - \hat{Y}]^2$) of the observed (Y) and predicted ($\hat{Y}$) values of the response variable. This minimization requirement is known as the *least squares criterion*.

The correlation coefficient can be computed for X and Y in regression analysis. It is, technically, only a measure of the "goodness-of-fit" of the regression line to the X, Y data, but it is in this capacity that the correlation coefficient describes the degree of empirical association between X and Y.

CORRELATION COEFFICIENT

Correlation analysis

The product-moment correlation coefficient (r) for two variables X and Y is computed from the following formula (Sokal and Rohlf, 1969, p. 498–508; and Draper and Smith, 1981, p. 43–45):

$$r = \frac{\sum_{i=1}^{N} ([Y_i - \bar{Y}] \cdot [X_i - \bar{X}])}{\sqrt{\sum_{i=1}^{N} (Y_i - \bar{Y})^2 \cdot \sum_{i=1}^{N} (X_i - \bar{X})^2}} \tag{1}$$

where $\bar{Y}$ and $\bar{X}$ are the means of the N observed Y_i and X_i values, respectively. Both X and Y are assumed to contain random error. This formula is a reduced form of the ratio of the covariance of X and Y to the product of the standard deviations of X and Y.

The product-moment correlation coefficient can describe only the extent of linear association between two variables. Its value ranges between -1 (perfect inverse linear association; one variable increases as the other decreases with all X, Y sample points falling on a straight line) and $+1$ (perfect direct linear association; both variables increase together with all X, Y sample points again falling on a straight line). A correlation of zero indicates a complete absence of linear association with the X, Y sample points exhibiting a random scatter when plotted. Values intermediate between 0 and ± 1 indicate various degrees of linear association.

The product-moment correlation coefficient can also be used to describe the strength of curvilinear relationships, but to do this it is necessary to first "linearize" the data. Linearization can often be achieved by applying a mathematical transformation to one or both of the variables (Acton, 1959, p. 219–223; and Draper and Smith, 1981, p. 218–225). For example, if Y varies logarithmically with X, the scatter plot of Y versus log X will have a linear trend and the correlation coefficient will consequently indicate a strong linear association between Y and log X.

The necessity of linearizing curvilinear relationships, which are common in geology, is a major drawback to correlation analysis. The problem is that the data analyst must have an *a priori* knowledge of the mathematical relationship between X and Y. The number of possible mathematical functions is infinite, and so what typically happens is that the data analyst settles for a relatively simple and well-known function that seems to do a good job of linearizing the X, Y relationship.

Regression analysis

The regression approach to association analysis involves finding a regression line that closely approximates the trend of the X, Y sample points in a scatter plot. One does this by specifying the general form of the polynomial equation (i.e., regression model) for a straight or curved line, and then "fitting" the equation to the data. The fitting process consists of finding optimal values for the unknown parameters in the equation through the application of least squares regression (Crow et al., 1960, p. 147–167; Davis, 1973, p. 192–221; and Draper and Smith, 1981, p. 5–17). The equation of the best-fit line is the prediction equation for the "regression of Y on X."

Only three polynomial regression models are generally of interest in bivariate association analysis (Table 9.1). The first degree polynomial is the equation of a straight line and thus can be used to represent linear trends. The second degree polynomial is the equation of a parabola whose axis of symmetry is parallel to the Y axis. It can be used to approximate simple curvilinear trends. The third degree polynomial is the equation of

TABLE 9.1 Polynomial Regression Models

Polynomial degree	Model[a,b]	Prediction equation[c]
First	$Y_i = a + (b_1 \cdot X_i) + \epsilon_i$	$\hat{Y}_i = a + (b_1 \cdot X_i)$
Second	$Y_i = a + (b_1 \cdot X_i) + (b_2 \cdot X_i^2) + \epsilon_i$	$\hat{Y}_i = a + (b_1 \cdot X_i) + (b_2 \cdot X_i^2)$
Third	$Y_i = a + (b_1 \cdot X_i) + (b_2 \cdot X_i^2) + (b_3 \cdot X_i^3) + \epsilon_i$	$\hat{Y}_i = a + (b_1 \cdot X_i) + (b_2 \cdot X_i^2) + (b_3 \cdot X_i^3)$

[a] Y_i and X_i are the response and predictor variables, respectively, for the ith sample; ϵ_i is the random error associated with the ith sample; a is the Y-axis intercept coefficient; and b_1, b_2, and b_3 are regression coefficients.

[b] $(b_1 \cdot X_i)$, $(b_2 \cdot X_i^2)$, and $(b_3 \cdot X_i^3)$ are referred to as the linear, quadratic, and cubic regression terms, respectively.

[c] $\hat{Y}_i$ is the predicted value of the response variable for the ith sample ($\hat{Y}_i = Y_i - \epsilon_i$). The values of the coefficients are obtained through least squares regression analysis.

a double parabola whose two axes of symmetry also parallel the Y axis. It can be used to approximate low order, complex curvilinear trends. The vast majority of bivariate trends can be represented adequately by just the first and second degree polynomials. The second and third degree polynomials are highly flexible in the sense that they can closely mimic a wide variety of mathematical functions. In most cases, only a segment of the polynomial curve is actually fitted to the data: just that portion of the curve that comes closest to matching the trend. The only kinds of curvilinear trends that do not respond well to polynomial curve fitting are those with long, straight segments, or reversals in the X values.

A statistic known as the multiple correlation coefficient, R (or, for R^2, the coefficient of multiple determination), is a measure of the goodness-of-fit of the regression line to the X, Y data (Davis, 1973, p. 197; and Draper and Smith, 1981, p. 33, 90):

$$R = \sqrt{\frac{\sum_{i=1}^{N} (\hat{Y}_i - \overline{Y})^2}{\sum_{i=1}^{N} (Y_i - \overline{Y})^2}} \qquad (2)$$

where $\hat{Y}_i$, Y_i, $\overline{Y}$, and N are as previously defined (equation 1 and Table 9.1). The term "multiple" alludes to the widespread use of this statistic in multivariate regression where there are multiple predictor variables. The multiple correlations associated with the first, second, and third degree polynomials are referred to as the linear, quadratic, and cubic correlations, respectively. R ranges in value from $+1$ (perfect association; all sample points fall on the regression line) to zero (no association; sample points exhibit random scatter). Varying degrees of association are indicated by values between these two extremes. It is interesting to note that, for the special case of linear regression, the multiple correlation is equal to the product-moment correlation between X and Y; and that, for any polynomial regression model, the multiple correlation is equal to the product-moment correlation between $\hat{Y}$ and Y (Draper and Smith, 1981, p. 33, 46).

The coefficient of multiple determination is a statistically more meaningful parameter than the multiple correlation coefficient because it represents the proportion of the total variation in Y that is accounted for by the regression line. A more heuristic explanation of the meaning of R^2 is that it is a parameter proportional to the amount of scatter in the sample points about the regression line ($Y - \hat{Y}$) and thus is a measure of the strength of bivariate association.

As has been previously stated, regression analysis is, strictly speaking, appropriate only for experimental associations where the response variable (Y) is subject to random error and the values of the predictor variable (X)

are exact. It frequently happens, however, that the X values in experimental associations also contain some random error. This is to be expected because, although set by the investigator, the X values are still measured quantities and, therefore, are subject to the reproducibility limits of the measuring process. As long as the error in X remains small in comparison to the experimental variation in $X,$ the regression analysis will remain valid (Draper and Smith, 1981, p. 123–124).

EXPLORATORY ASSOCIATION ANALYSIS

Regression analysis can be and often is applied to natural bivariate associations, the most common type in geological data sets. There are compelling reasons for doing so. Correlation analysis is limited to describing the extent of linear association. Only through the often difficult and inaccurate process of linearization can it be applied to curvilinear relationships. Regression analysis is far more convenient because polynomial curves can be easily fitted, with reasonable accuracy, to almost any curvilinear trend. It is true, of course, that for natural associations the best-fit regression equation will not be optimal in the sense of minimizing all the random error in the data. It minimizes only the error in the response variable and none in the predictor variable. Despite this, regression analysis, with its associated multiple correlation coefficient, still provides an adequate description of the strength of a curvilinear relationship in natural associations. Correlation analysis does remain the preferred procedure for those situations where a bivariate relationship is strictly linear or can be reliably linearized. It is important to recognize, however, that in the former case the product-moment and multiple correlations are equivalent.

Geological data sets often contain numerous variables. The extent of bivariate association is usually of interest. A reasonable approach to the exploratory treatment of such data is to compute the multiple linear, quadratic, and cubic correlation coefficients (obtained through regression analysis) for those variable pairs of interest. This can be done without regard to whether the bivariate associations are natural or experimental. The procedures described in later sections of this chapter will aid the data analyst in selecting the most appropriate correlation coefficient for a given variable pair.

If, as recommended above, regression analysis is to be applied indiscriminately to natural and experimental associations, it is necessary to have guidelines governing the designation of the response and predictor variables for quadratic and cubic regressions. The multiple correlation coefficient for linear regression will be the same regardless of which variable is treated as the predictor. For experimental associations, the predictor variable will always, of course, be the variable that was experimentally manipulated. For natural associations, the choice of predictor variable would be based on the following considerations:

1. An apparent cause-and-effect relationship exists. The predictor variable would be the one influencing, at least potentially, the values of the other (response) variable. An example would be permeability and porosity in sandstones. Porosity would be designated the predictor variable because it is well known that permeability is a derived rock property that is, in part, dependent on porosity.

2. No cause-and-effect relationship exists, or there is uncertainty as to its existence. The predictor variable would be the one that:

   a. Does not exhibit a reversal of values as the other variable increases;

   b. Has the least measurement error (the sampling error would be the same for both variables because the corresponding "paired" values of each are measured on the same sample); or

   c. Provides the highest quadratic correlation for the two regression analyses where each variable is alternately treated as predictor and response variables (see the next section for justification).

   Of the three decision rules, the first should take precedence followed by the second and then the third. An example of a natural association that lacks an obvious cause-and-effect relationship is porosity and tortuosity in sandstones (Figure 9.1). The first two decision rules offer no guidance. Inspection of Figure 9.1 reveals that no reversal in trend will occur for either variable when treated as the predictor. The measurement errors are the same for both variables because they were measured by point counting from the same thin sections. The two quadratic correlations are 0.723 (for porosity as predictor) and 0.880 (for tortuosity as predictor). Applying the third decision rule, the functional designation that maximizes the strength of the bivariate relationship (i.e., provides the highest quadratic correlation) is the one where tortuosity is treated as the predictor variable. An example of two variables that differ in their measurement errors would be volume percent alkali feldspar (from a thin-section point count) and weight percent rubidium (from a whole-rock atomic absorption analysis) in granite. The second decision rule would designate rubidium the predictor variable because whole-rock analyses are more reproducible than thin-section analyses.

ADVANTAGES OF THE QUADRATIC CORRELATION COEFFICIENT

Data analysts, as noted earlier, seem to have a propensity for describing bivariate relationships with a linear correlation coefficient. When this statistic is computed for data with a curvilinear trend, the strength of the bivariate relationship will be underestimated. Any subsequent statistical analyses (e.g., R-mode cluster or principal components analyses) or geologic interpretations that are based on the misapplication of linear correlations may lead ultimately to erroneous conclusions. This problem suggests that the data analyst should compute both linear and quadratic (and possibly cubic) correlation coefficients and, by some means, then decide which is most appropriate for a given pair of variables. Such an approach would, of course, require more time and effort. It would also be unnecessary because the quadratic correlation coefficient adequately describes the strength of *both* linear and simple curvilinear trends.

If one were to compute linear and quadratic correlation coefficients for a data set with a linear trend, the quadratic correlation would be equal to or negligibly larger than the linear correlation. This is to be expected because the addition of the quadratic term $(b_2 \cdot X_i^2)$ to the regression model would contribute essentially nothing to the regression of Y on X and, thus, there would be no significant improvement in fit or correlation going from the first to the second degree polynomial (Sokal and Rohlf, 1969, p. 540; and Draper and Smith, 1981, p. 91). Clearly, the quadratic correlation coefficient describes the strength of both linear and simple curvilinear trends. If one were intent on relying on a single correlation coefficient, as is often the case with the linear correlation in geology, it would be much better to use the quadratic correlation.

STATISTICAL INFERENCE

Real vs. spurious relationships

If one were to compute the linear (or any other) correlation coefficient for a set of X, Y data values where X and Y were totally unrelated to one another, one would find that it would not equal zero. An example of such a data set would be $Y =$ annual number of earthquakes globally with Richter magnitudes over 5.5, and $X =$ annual enrollment of undergraduate geology majors in U.S. universities. The linear correlation for X and Y would certainly be very low, but it probably would not be zero. In such a case, the data exhibits a spurious correlation. This type of correlation is due to chance covariation between unrelated X and Y values (see Mann, Chapter 6).

Every observed correlation contains potentially spurious and real parts. The real part would result from either a cause-and-effect or mutual dependence relationship between X and Y. The spurious part has a complex origin. The ith observed data values for variables X and Y (X_i and Y_i) can be represented as follows: $X_i = x_i + \epsilon_{x,i}$ and $Y_i = y_i + \epsilon_{y,i}$ where x_i and y_i are the "true" values, and $\epsilon_{x,i}$ and $\epsilon_{y,i}$ are the random errors associated with X_i

and Y_i, respectively. Even if no real relationship exists between X and Y (i.e., the correlation between X and Y in the sampled population is zero), it is still possible to have a nonzero, spurious correlation between observed X and Y values because of the chance covariation between x and y, x and ϵ_y, y and ϵ_x, and/or ϵ_x and ϵ_y. If, on the other hand, a real relationship exists between X and Y, then the spurious part of an observed correlation can be attributed to the chance covariation for only the last three pairings. The spurious part of a correlation tends to increase with decreasing sample size because the opportunity for chance covariations increases as the number of paired observations decreases. The term "spurious correlation" is used in this chapter to refer to all sources of chance covariation and not just that between x and y, as is frequently the case in the statistical literature.

Data analysts are usually interested in applying the results of an association analysis to the population (either natural or experimental) from which the samples were drawn. The question to be answered is: "To what extent can the observed correlation for a sample be considered spurious?" If the result of a test of statistical significance indicates that it is probably largely spurious then it can be concluded that there is no significant relationship between X and Y in the sampled population.

If, as advocated in this chapter, the data analyst employs the regression approach to exploratory association analysis, it then becomes possible to test the statistical significance of the multiple linear, quadratic, and cubic correlations by both direct (F test of correlation coefficients) and indirect (F test of regression coefficients) means. It must be kept in mind, however, the traditional statistical tests, like the F test, are exact for experimental associations but only approximate for natural associations. A new "randomization test," appropriate for both experimental and natural associations, will be introduced in a later section.

Underlying assumptions

In order for tests of statistical significance to be valid, certain assumptions about the samples and population must be true. The assumptions listed below are those required by traditional tests that use the F statistic.

1. The set of samples is representative of the population from which it was drawn. A large number of systematic or, better yet, random samples is required to satisfy this assumption. Good discussions of systematic and random sampling strategies can be found in Krumbein and Graybill (1965, p. 147–169), Koch and Link (1970, p. 43–78), and Size (Chapter 1).

2. The sample contains no "accidental" errors. These are sampling and measurement errors incurred through carelessness or ineptitude of the investigator (e.g., sampling the wrong rock unit, misuse of analytical equipment, using contaminated samples, combining data from different

sources where analytical techniques were not identical, etc.) or mal-functioning (i.e., inaccuracy) of analytical equipment.

3. **a.** Random error exists only in the response variable (Y) whereas the values of the predictor variable (X) are error-free. The random error in Y has two sources: sampling and measurement. The sampling error arises from the inherent variability of Y in the sampled population and the measurement error arises from the reproducibility (i.e., precision) of the measurement process. For each value of X, there exists a subpopulation of Y values. In general, the single observed value of Y for a given X will not correspond to the mean of the Y subpopulation, and the extent to which it does not is a function of the random error.

    b. The Y subpopulations for the X values are all normally distributed and have the same variance (i.e., exhibit homoscedasticity).

    c. The random error associated with each observed Y value is unaffected by (i.e., is statistically independent of) the magnitude of the random errors associated with the other $N - 1$ observed Y values.

In geological studies, seldom will all, or even most, of these assumptions be true. In such cases, the results of a significance test and, in particular, the probability statement associated with it are not to be taken literally. When underlying assumptions are violated, the significance of the results is less, often considerably less, than the test would seem to indicate. It is probably true, however, that the correct interpretation of an observed correlation is more likely with the test results in hand, provided that the data analyst recognizes the limitations of those results.

Traditional significance tests

The statistical significance of linear, quadratic, and cubic correlations can be evaluated using a test based on the F statistic (Crow et al., 1960, p. 159, 178, and 241). The null (H_0) and alternative (H_a) hypotheses for this test are:

$H_0: r_m = 0$, a real mth degree polynomial relationship between X and Y does not exist in the sampled population (i.e., the observed nonzero sample correlation is entirely spurious).

$H_a: r_m \neq 0$, a real mth degree polynomial relationship between X and Y does exist in the sampled population (i.e., at least part of the observed sample correlation is real).

This test can be applied to any of the three correlation coefficients, but it is appropriate only for the linear correlation as is evident from the following example. If first and second degree polynomials are fitted to data with a strong linear trend, both the linear and quadratic correlations will

TABLE 9.2 Interpreting the Results of the F Test

Significant[a] regression terms	Significant correlations		
	Linear	Quadratic	Cubic
$b_1 \cdot X$	Yes[b]	No	No
$b_2 \cdot X^2$	No	Yes	No
$b_1 \cdot X$ and $b_2 \cdot X^2$	No	Yes	No
$b_3 \cdot X^3$	No	No	Yes
$b_1 \cdot X$ and $b_3 \cdot X^3$	No	No	Yes
$b_2 \cdot X^2$ and $b_3 \cdot X^3$	No	No	Yes
$b_1 \cdot X$, $b_2 \cdot X^2$, and $b_3 \cdot X^3$	No	No	Yes

[a]Significant at some specified level (e.g., 95% or 99%).

[b]The F test for the linear regression term is equivalent to the F test of the linear correlation coefficient.

pass the significance test. The test results are misleading because the quadratic term, $b_2 \cdot X^2$, contributes virtually nothing to the regression of Y on X (b_2 would have a value near zero). The linear term, $b_1 \cdot X$, is responsible for both the large linear and quadratic correlations (the two correlations will be nearly identical because of the lack of contribution from the quadratic term).

Because of the problems associated with testing the correlation coefficients directly, it is more useful to test the statistical significance of the individual regression coefficients (b_1, b_2 and b_3). This test, which also utilizes the F statistic, allows us to evaluate the significance of the contribution of the linear ($b_1 \cdot X$), quadratic ($b_2 \cdot X^2$), and cubic ($b_3 \cdot X^3$) regression terms to the explanation of the variation in Y (Davis, 1973, p. 215–217; and Draper and Smith, 1981, p. 31–33). An equivalent test could be applied that utilizes the t statistic (Crow et al., 1960, p. 160–161 and 179–181). From the results of the F test, the significance of the correlations can be inferred (Table 9.2).

The null (H_0) and alternative (H_a) hypotheses for the test are:

$H_0: b_k = 0$, the kth regression coefficient equals zero in the sampled population (i.e., the contribution of the kth regression term is entirely spurious).

$H_a: b_k \neq 0$, the kth regression coefficient does not equal zero in the sampled population (i.e., the contribution of the kth regression term is, at least in part, real).

The calculation procedure (Davis, 1973, p. 215–217) is as follows:

Step 1 Fit the first, second and third degree polynomials to the X, Y data. Obtain the sum of squares due to regression [$\Sigma(\hat{Y} - \overline{Y})^2$; SSR] for each of the three regressions—SSR_1, SSR_2, and SSR_3, respectively.

Step 2 Compute the contribution (as measured by *SSR*) of each regression term.

Term	Contribution		Calculation
$b_1 \cdot X$	SSR_L	=	SSR_1
$b_2 \cdot X^2$	SSR_Q	=	$SSR_2 - SSR_1$
$b_3 \cdot X^3$	SSR_C	=	$SSR_3 - SSR_2$

Step 3 Compute the mean sum of squares due to deviations for the third degree polynomial (MSD_3). MSD_3 is the sum of squares due to deviations for the third degree polynomial [$\Sigma(Y - \hat{Y})^2$; SSD_3] divided by its degrees of freedom (N − 4).

Step 4 Compute the F statistic for each regression term.

Term	Calculation
$b_1 \cdot X$	$F_L = SSR_L/MSD_3$
$b_2 \cdot X^2$	$F_Q = SSR_Q/MSD_3$
$b_3 \cdot X^3$	$F_C = SSR_C/MSD_3$

Step 5 Obtain the critical (e.g., 95% where the alpha probability, α, equals 0.05) F value from a table of the probability distribution of the F statistic (Crow et al., 1960, Table 5, p. 234–239; Draper and Smith, 1981, p. 533–535; and many other statistics books). Degrees of freedom for this table are 1 (column) and $N - 4$ (row). Compare the computed F statistic for each regression term against this critical F value. If the statistic is larger than the critical value, one rejects the null hypothesis and accepts the alternative hypothesis. One concludes, if $\alpha = 0.05$, that only 5% of the time will the computed F statistic be larger than the critical F value when the null hypothesis is true.

The F test for the regression coefficients is, strictly speaking, valid only if all the previously described underlying assumptions are true. As already pointed out, this is unlikely to be the case. Assumption 3 is the most unreasonable from a geological standpoint and thus is the most likely to be violated. If assumptions 1 and 2 can be accepted, then the statistical significance of the regression coefficients can be evaluated using random number experiments in what is generally referred to as a randomization test (Sokal and Rohlf, 1969, p. 629–630). This test allows one to by-pass assumption 3. We need not be concerned with the exact nature of the random error in Y or X because its characteristics are incorporated, to a certain extent, into the empirical distributions for Y and X. A randomization test can only yield approximate results, but the results would seem to be more applicable to natural associations and nonideal experimental associations than the results of the more restrictive F test.

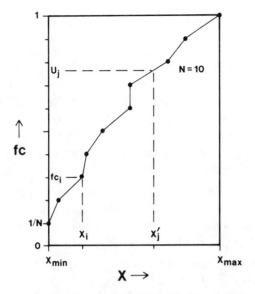

Figure 9.3 Cumulative frequency curve for the empirical sample distribution of X.

Randomization test

In the randomization test, regression analysis is performed on bivariate sets of random numbers. Each set consists of N paired but unrelated values. The distributions of values for the two random number variates in a set are the same as the observed X and Y sample distributions. From the regression results, one is able to estimate the chance (spurious) covariation between Y and X, X^2 and X^3.

The calculation procedure is as follows:

Step 1 Generate N values for a random number variate, X', that has the same distribution as X. Use the approach of Harbaugh and Bonham–Carter (1970, p. 74–78).

    i. Arrange the N X values in order of increasing magnitude.

    ii. Plot the distribution of ordered X values using a cumulative frequency curve (Figure 9.3). The cumulative frequency (fc) for the ith X value (fc_i) is equal to the number of X values smaller than or equal to (but preceding) X_i divided by N, the total number of X values ($i = 1, \ldots, N$).

    iii. Generate N "uniform" random numbers between $1/N$ and 1 (U_j, $j = 1, \ldots, N$). These numbers are randomly drawn from a distribution where every number occurs with equal frequency between $1/N$ and 1.

iv. Find the value on the X axis corresponding to each random number. For the jth uniform random number the value of X is X'_j. *Note:* The distribution of the N X' values will be similar (or nearly identical if N is very large) to the empirical distribution of X. The cumulative frequency curve is steeper where the observed X values are more frequent and, thus, the random numbers and their corresponding X' values also are necessarily more frequent for steeper segments of the curve.

Step 2 Generate N values for a random number variate, Y', that has the same distribution as Y (follow the procedure in Step 1).

Step 3 Combine X' and Y' subsets to obtain a data set of N paired but unrelated values. This set of values represents the sampled population (if assumptions 1 and 2 are valid) when the null hypothesis ($r_m = 0$ or $b_k = 0$) is true.

Step 4 Compute F statistics for the linear, quadratic, and cubic regression terms for the X', Y' random number data set (follow the procedure described earlier—Steps 1 through 4—for the F test).

Step 5 Repeat Steps 1 to 4 a large number of times (p), each time generating a different X', Y' random number data set. The larger p is, the more reliable the test results. A minimum value for p would perhaps be one or two hundred.

Step 6 Take the p F_L, p F_Q, and p F_C values for the random number data sets and separately arrange them in order of increasing magnitude. These ordered sets of F values constitute the probability distributions for the spurious F statistics (Figure 9.4).

Figure 9.4 Probability distribution of the spurious F statistic for the kth regression term.

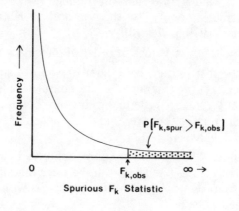

Step 7 Compute F statistics for the linear, quadratic, and cubic regression terms for the observed X, Y sample data.

Step 8 To assess the statistical significance of the kth regression term, one need only determine the percentage of spurious F_k values that are smaller than the single observed F_k value ($F_{k,\text{obs}}$). If the percentage was 89, for example, then b_k would be declared significant at the 89% level (i.e., there is only an 11% chance of obtaining a spurious F value larger than the observed one when the null hypothesis, $H_0 : b_k = 0$, is true; $P\{F_{k,\text{spur}} > F_{k,\text{obs}}\}$ in Figure 9.4). The data analyst must decide what constitutes a sufficiently large significance level in order for the contribution to be considered "real." Most statisticians would recommend the 95% level.

Comparison of the randomization and *F* tests

Table 9.3 compares the significance levels obtained from the randomization and F tests for some selected association analysis results. The significance levels for the F test were not taken from a table of critical F values but rather were calculated using the algorithm of Dorrerr (1968). It is apparent from this table that there is close agreement between the two sets of significance levels. This agreement is further substantiated by Figure 9.5. The results of the randomization test would seem to be more reliable but those of the F test are sufficiently similar so that it really does not matter which set of results is used. Figure 9.5 illustrates what has long been claimed but never adequately demonstrated by statisticians: the F test is robust in that it is not greatly affected by violations of its underlying assumptions.

An obvious advantage to using the F test in place of the randomization test is that the latter requires specialized computer software (available from the author) and, for some computers, a prohibitively long execution time. The F test is more convenient because of readily available tables with critical F values and computer software for calculating F statistics.

VERBAL SCALE FOR THE STRENGTH OF A CORRELATION

Even if one accepts the results of a significance test and concludes that a real relationship does exist between a pair of variables, the strength of the relationship still has not been established. The relationship could exist and yet be weak. It is conventional to consider correlations over 0.80 as indicative of a "strong" relationship. This is a dangerous practice because for a given correlation the "apparent" strength of the relationship may be a function of the sample size. A 0.80 linear correlation for 100 samples

TABLE 9.3 Significance Test Results

| Variables[a] | Coefficients | Observed correlation | Significance levels | | Significant correlation[c] |
			Randomization test[b]	F test	
Kurtosis (Y) vs. Mean (X)	Linear	−0.510	99	99	Linear
	Quadratic	0.518	39	37	
	Cubic	0.527	45	42	
Kurtosis (Y) vs. Skewness (X)	Linear	0.585	99	99	Quadratic
	Quadratic	0.671	96	96	
	Cubic	0.690	69	71	
Mean (Y) vs. Sorting (X)	Linear	−0.024	8	10	Quadratic
	Quadratic	0.378	96	94	
	Cubic	0.412	69	61	
Mean (Y) vs. Skewness (X)	Linear	0.062	35	28	Cubic
	Quadratic	0.392	96	96	
	Cubic	0.562	97	97	
Skewness (Y) vs. Sorting (X)	Linear	0.235	76	77	None
	Quadratic	0.302	72	66	
	Cubic	0.356	72	66	

[a]Moment grain size statistics for 28 sand samples (Harrell, 1983): Mean grain size (Mean), standard deviation of grain size (sorting), and skewness and kurtosis of the grain size distribution. Y indicates the response variable and X, the predictor variable.

[b]Based on 200 random number data sets per test ($p = 200$).

[c]Based on the interpretations in Table 9.2 and a 95% threshold for significance.

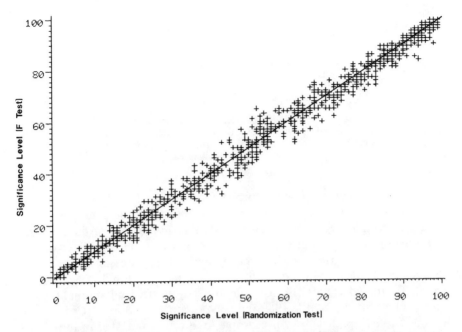

Figure 9.5 Comparison of significance levels for the randomization and F tests. The number of paired values is 1140. These values were obtained from significance tests for the linear, quadratic, and cubic regression terms for all possible pairs of 20 petrologic variables (including those in Table 9.3) in a data set consisting of 28 sand samples (Harrell, 1983). Two hundred random number data sets were used for each randomization test. The diagonal line represents the best-fit straight line ($Y = -0.171 + 0.998 \cdot X; r = 0.996$).

should be regarded more favorably than the same correlation for 10 samples (see Mann, Chapter 6). This is true because the spurious part of a sample correlation tends to increase with decreasing sample size. What is needed is a standardized verbal scale for describing the strength of bivariate relationships. The scale should take into account the sample size and, hence, the likely magnitude of the spurious correlation. Separate scales would have to be devised for the linear, quadratic, and cubic correlations.

The verbal scale proposed here is dependent upon the estimated magnitude (i.e., a value corresponding to a selected percentile of the probability distribution) of the spurious correlation (r_s) for a given sample size and polynomial degree (Figure 9.6). The spurious correlation can be computed from the following formula (modified from Harris, 1975, p. 19):

$$r_s = \sqrt{\dfrac{\dfrac{m \cdot F_s}{N - m - 1}}{1 + \dfrac{m \cdot F_s}{N - m - 1}}}$$

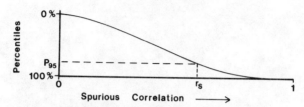

Figure 9.6 Probability distribution of the spurious correlation for a given sample size and polynomial degree. P_{95} is the percentile corresponding to the probability of occurrence (95%) of a spurious correlation less than r_s.

where F_s is the spurious F value corresponding to a selected percentile (e.g., 95%) of the probability distribution of the F statistic for a sample size of N, a polynomial degree of m, and a true null hypothesis of no bivariate relationship in the sampled population. F_s is equivalent to the critical F value discussed earlier and can be obtained from the same table. The degrees of freedom will, however, differ: m (column) and $N - m - 1$ (row). Tabulated values of r_s for different sample size, polynomial degree, and probability percentile (95% and 99%) can be found in Crow et al. (1960, Table 7, p. 241). If, for example, one were interested in the verbal scale for the quadratic correlation ($m = 2$) and a sample size of $N = 28$, then the 95th percentile values of F_s and r_s would be 3.39 and 0.462, respectively.

The verbal scale for a given correlation is constructed by subdividing the difference for $1 - r_s$ into five equal intervals:

Interval		*Correlation Strength*
$(r_s + [4 \cdot I])$ to	1	Very strong
$(r_s + [3 \cdot I])$ to	$(r_s + [4 \cdot I])$	Strong
$(r_s + [2 \cdot I])$ to	$(r_s + [3 \cdot I])$	Moderate
$(r_s + I)$ to	$(r_s + [2 \cdot I])$	Weak
r_s to	$(r_s + I)$	Very weak
Less than r_s		Null

where $I = (1 - r_s)/5$. The verbal scale for the quadratic correlations in Table 9.3 is given in Table 9.4. Applying the scale, correlation strengths for variable pairs in Table 9.3 are, from top to bottom: very weak, weak, null, null and null. An apparent discrepancy exists between the strength of the quadratic correlations and the statistical significance of the quadratic regression terms. This should not be surprising because the verbal scale considers the overall strength of the bivariate relationship whereas the test results pertain only to a single regression term.

TABLE 9.4   Verbal Scale for the
Quadratic Correlations in Table 9.3

Interval[a]	Correlation strength
0.893 to 1	Very strong
0.785 to 0.892	Strong
0.677 to 0.784	Moderate
0.570 to 0.676	Weak
0.462 to 0.569	Very weak
Less than 0.462	Null

[a]Based on a probability percentile of 95%, a sample size of 28, and a polynomial degree of 2.

The spurious correlation on which the verbal scale is based can be used to test the statistical significance of an observed sample correlation. If the sample correlation is larger than r_s, then it can be considered significant (i.e., the null hypothesis of zero correlation in the sampled population is rejected). Such a test, as pointed out earlier, is valid only for the linear correlation. Despite this, the verbal scale is still meaningful for the quadratic (or cubic) correlation because it takes into account the likely magnitude of the "spurious part" of the observed correlation. The scale provides the data analyst with an objective and reproducible means of characterizing the overall strength of a bivariate relationship.

CLOSING REMARKS

The principal problem faced by practitioners of association analysis is violation of the underlying assumptions for tests of statistical significance. It is to be expected that some of these assumptions will be violated by most data sets. The randomization and F tests have been shown to be largely immune to violations of the third assumption, but no test of significance can overcome violations of the first and second assumptions. If a sample is not representative of the population or contains accidental error, no meaningful interpretation of the data is possible. Accidental error, by its nature, can easily be avoided by exercising care in the sampling and analytical procedures, and a statistically sound sampling design can help to assure accurate representation of a population. Also, data that are demonstratively "bad" can be removed from a data set.

The analysis of bivariate associations is further complicated by the use of ratio variables (e.g., MgO vs. FeO/[MgO + FeO]) and variables of a constant sum (e.g., SiO_2 vs. Al_2O_3 for oxides summing to 100%). In both cases the variable pairs are not mathematically independent of one another and so possess a "built-in" or "induced" linear correlation. To determine the statistical significance of the observed correlation between ratio variables or variables of a constant sum, it is necessary to estimate the likely magnitudes of both the spurious and induced correlations. There is, at present, no entirely satisfactory way of doing this. The problem has been discussed by Chayes (1960, 1971) and Chayes and Kruskal (1966). One possible approach would be to use random number experiments similar to those employed in the randomization test.

The interpretation of bivariate associations is not always a straightforward proposition. It requires a data analyst to dissect an observed correlation into its real and spurious (and, sometimes, induced) parts. Tests of statistical significance are used for this purpose but the usefulness of these tests depends on the validity of their underlying assumptions. It will sometimes be the case that one or more of these assumptions are violated unbeknownst to the data analyst. Use of a test is still worthwhile because it provides an objective and reproducible means of analyzing data. Data analysts would be far worse off if they had to rely instead on a subjective evaluation of sample point patterns in scatter plots. Data analysts who use tests of statistical significance are obliged to be more conservative (i.e., less certain) in their conclusions than would seem to be necessary from the computed significance levels.

REFERENCES

Acton, F. S., 1959, Analysis of straight-line data: Dover, New York, 267 p.

Chayes, F., 1960, On correlations between variables of constant sum: *Jour. Geophysical Research,* v. 65, p. 4185–4193.

Chayes, F., 1971, *Ratio correlation—A manual for students of petrology and geochemistry:* Univ. Chicago Press, Chicago, 99 p.

Chayes, F., and W. Kruskal, 1966. An approximate statistical test for correlations between proportions: *Jour. Geology,* v. 74, p. 692–702.

Crow, E. L., F. A. Davis and M. W. Maxfield, 1960, *Statistics manual:* Dover, New York, 288 p.

Davis, J. C., 1973, *Statistics and data analysis in geology:* Wiley, New York, 550 p.

Dorrerr, E., 1968, *Algorithm 322—F distribution:* Assoc. Computing Machinery Communications, v. 11, p. 116–117.

Draper, N. R., and H. Smith, 1981, *Applied regression analysis:* Wiley, New York, 709 p.

Harbaugh, J. W., and G. Bonham–Carter, 1970, *Computer simulation in geology:* Wiley, New York, 575 p.

Harrell, J., 1983, Grain size and shape distributions, grain packing, and pore geometry within sand laminae—Characterization and methodologies: Ph.D. Dissertation, Univ. Cincinnati, Cincinnati, Ohio, 583 p.

Harris, R. J., 1975, *A primer of multivariate statistics:* Academic Press, New York, 332 p.

Koch, G. S., and R. F. Link, 1970, *Statistical analysis of geological data,* vol. 1: Wiley, New York, 375 p.

Krumbein, W. C., and F. A. Graybill, 1965, *An introduction to statistical models in geology:* McGraw-Hill, New York, 475 p.

Sokal, R. R., and F. J. Rohlf, 1969, *Biometry:* Freeman, San Francisco, 776 p.

Index